Vending Machine Business

Business

A Simple Guide to Start a Vending Machine

(Grow and Promote a Profitable Vending Machine Business)

Adan Jordan

Published By **Bella Frost**

Adan Jordan

All Rights Reserved

Vending Machine Business: A Simple Guide to Start a Vending Machine (Grow and Promote a Profitable Vending Machine Business)

ISBN   978-1-77485-716-8

No part of this guidebook shall be reproduced in any form without permission in writing from the publisher except in the case of brief quotations embodied in critical articles or reviews.

Legal & Disclaimer

TABLE OF CONTENTS

Introduction

Everyone in every town has their favorite bar, but how often do you see them selling snacks? You might not see vending machines at bars, but they're a great way to keep your customers' stomachs happy, and if you want to start your own business selling snacks, it's the perfect opportunity. These days the vending machine market is on the rise - some machines will make money for you too. If you're interested in starting a new venture or getting into this specific market, here is an introduction on how to start a vending machine business.

The first thing to decide is what type of snacks you're going to sell. Generally, the most profitable items are candy, chips, and energy bars, but there are other ways to profit from the machine. The vending machines that make money for you can be found in local stores, like pharmacies or convenience stores (CVS). You can also find these machines online through websites that sell snacks online.

These snack vending machines will often have 30 different snacks, but some may have up to 100. They'll usually have a price point of $3 a bag or more. You'll need at least six bags of snacks each week to make your business profitable enough, and this is highly dependent on what you're selling. You'll also want to see what your competition is doing when looking for machines to place in different locations. Talk to local distributors, managers of the stores where you want to set up shop, etc.

Once you've decided on the snacks you want to sell, find a location for your machine. You'll need a permit from the Department of Health. They will require that you clean and sanitize your machine regularly since you'll be getting many people from different cities touching it every day. You will also need a permit from the Department of Agriculture since they're the ones who regulate vending machines. Also, make sure that there isn't a ban on certain snacks or products in the city.

Place the machine in an area you think will do well with your product. If you're selling premium, higher-priced snacks, it may be best to place them in a high-traffic area close to offices or where many people are working. But if you want to sell lower-priced snacks, try putting them near a play area or bathrooms - places that attract kids and families. Make sure that your products are easily accessible; consumers will get fed up and leave if they're too hard to reach.

If you have one location, check that the same items aren't always being sold before buying more. It's also a good idea to make sure items aren't damaged, wrapped incorrectly, etc... You may want to take some photos of the inventory and keep them on file if anything is ever stolen. If one of your snacks is being sold often enough, you can add more of that item to the machine. Some vending machines will have debit/credit card capabilities; this could be a way for customers to pay for the snacks instead of putting money in the machines themselves. As long as the card has cash on it and no additional fees are involved, it's worth checking out.

Try and stay up to date with current trends in snack food. If you're selling energy bars, be sure to look into the most popular flavors. If you were selling chips, be sure to offer different salsa flavors as well.

Don't get discouraged if your first week is a little dry. Don't buy too many bags of snacks because you can only bring in so much. Think about what you'll do next week and the following weeks to help build up your business; advertising and events can go a long way toward helping your business grow.

Vending machines are a constantly changing market, and you will need to adapt to it. Be ready to change your snacks based on what the public is requesting. If customers keep coming up to you saying they want "more chips," buy more. If you can't sell your snacks, you won't make any money.

Selling snacks out of vending machines is a career that anyone can get into. If you follow the steps above and look for the right machine for your business, you'll be on your way to success in no time. Just remember that solving problems related to vending machine sales are part of the experience, and if you try hard enough, there's a good chance you'll succeed!

Chapter 1: Who Is A Vendor?

A vendor is a person or business that supplies goods or services to a company. Another term for vendor is supplier.

In many situations a company presents the vendor with a purchase order stating the goods or services needed, the price, delivery date, and other terms. Generally, when the vendor delivers the goods or services it will also send an invoice to the company. The company should then match the vendor's invoice with its purchase order and receiving report. If the three documents are in agreement, the company can than arrange for payment to the vendor.

Qualities to Look for in your Vendor

No matter where you look, consumers have a clear need for financial services. Nearly 70 percent of adults globally have a bank account, according to the World Bank, and among developed countries, that figure is much higher at 94 percent. Technological innovation and omnichannel access have fueled increasing consumer demands, and branches nationwide have accelerated their efforts to fulfill them.

"How can you tell if your vendor is truly invested in your company's success?"

But banks and credit unions cannot and should not go it alone. Sure, you've spent years developing top-notch, personalized customer service but who is serving you when it comes to your critical financial technology, security and service needs? You need a vendor who is every bit as engaged in the customer experience and overall performance as you are, allowing your branch to rise to the top in an already crowded market.

When it comes to choosing a vendor partner, if only it were as easy to picking a name out of a phonebook: The truth is, not all vendors are created equal or have your key interests in mind.

The question then becomes: How can you tell if a vendor is truly invested in your success? Here are eight characteristics to look for when vetting technology and service partners:

1. Focus

Your primary focus are the people that come through your branch doors and log onto their online accounts. Are your vendors facilitating this goal? Are they focused on you and what you do or are they just focused on their profit margins and shareholders?

2. Responsiveness

Your customers don't want to wait for their needs to be addressed. Neither should you. Whether providing electronic security, automation, ATMs or any combination of solutions, a vendor should be ready, willing and available to work with you, in real time, to achieve to your short-term and long-term strategic goals.

3. Specialization

Run-of-the-mill, "one-size-fits-all" solutions no longer cut it. Your vendor should offer specialized solutions to serve your needs.

4. Results-driven

The process is only half the battle; quality vendors have to deliver on your expectations with the type of equipment and services that guarantee you'll get results. The best vendors have data and case studies that demonstrate their successes and support their effectiveness.

5. Fairness

The essence of a good vendor is captured in how it treats its clients, independent of their status or size. Whether its portfolio is made up of small-town community banks or global conglomerates, your vendor should be serving you with every bit as much verve and tenacity as larger entities â€" no playing favorites.

6. Customization

Banks may provide a litany of customer services all in one place, but every product and service is designed to meet a unique need. Vendor services should be customized to what you do, not the other way around.

7. Innovation
Customer preferences are never static, nor is technology. You can't rest on your laurels, so why would you allow your vendor to do so? They should be attuned to innovation and always looking to progress.

8. Investment
The only way to get better is by taking chances and laying it all on the line. You've got skin in the game and so should your vendors.
Contributions of Vendors

The Informal Economy Monitoring Study (IEMS) revealed ways in which street vendors in five cities strengthen their communities:

•   Most street vendors provide the main source of income for their households, bringing food to their families and paying school fees for their children.

- These informal workers have strong linkages to the formal economy. Over half the IEMS sample said they source the goods they sell from formal enterprises. Many customers work in formal jobs.
- Many vendors try to keep the streets clean and safe for their customers and provide them with friendly personal service.

- Street vendors create jobs, not only for themselves but for porters, security guards, transport operators, storage providers, and others.

- Many generate revenue for cities through payments for licenses and permits, fees and fines, and certain kinds of taxes. This was true of two thirds of street vendors in the IEMS sample.

- Street trade also adds vibrancy to urban life and in many places is considered a cornerstone of historical and cultural heritage. For example, street vendors who sell chai, called "chai-wallahs," are an important part of India's cultural heritage. See photos (NPR) and video of chai-wallahs in action. This Linked-In article explains how they are innovating this centuries-old practice to meet the demands of the present.

• Despite their contributions, street vendors face many challenges, are often overlooked as economic agents and unlike other businesses, are hindered rather than helped by municipal policies and practices.

Job Summary of a Vendor

Responsible for supplying specific goods or services to individuals or companies. May sell items at a fair, sporting event, market, or any place frequented by consumers.

Primary Responsibilities

• Supply goods to clients.
• Sell refreshments, programs, alcohol, novelties, or cushions at sports events, parades, concerts, or other venues.

- Takes orders and checks inventory to ensure products are in stock.
- Describes product features and tells people how to maximize its use.
- Hand, pass, or throw items to the purchaser and in turn receive payment, making change as necessary.
- Collect money for sales.
- Count money at the end of a shift and turn over income.
- Reimburse manager or boss for the provision of vended items.
- Search for purchasers among large crowds.
- Keep foodstuffs hot or cold.
- Dispense napkins, straws, and condiments as needed.
- Greet customers warmer and answer questions about products.
- Negotiate pricing for products and services.
- Replenish inventory as needed.
- Count out correct change using simple math.
- Manage different vendors.

- Clean up machines, bags, and other equipment at the end of a shift.
- Collect tips from customers.
- Ensure food meets health and safety standards and is kept at optimal temperatures and conditions.

Street Vending and Gender

In many countries, especially in Africa, the majority of street vendors are women: 88 per cent in Ghana, 68 per cent in South Africa, and 63 per cent in Kenya (ILO and WIEGO 2013). Only in a few countries where cultural norms restrict women's economic activities do women account for 10 per cent or less of street vendors.

Available evidence suggests a higher share of women than men sell perishable goods, which are more likely than other goods to spoil or to be confiscated. Other research has shown that women street vendors typically earn less than men and in many countries, less than half as much as men.

Street Vendor in Ghana

Low barriers to entry, limited start-up costs, and flexible hours are some of the factors that draw street vendors to the occupation. Many people enter street vending because they cannot find a job in the formal economy.

But surviving as a street vendor requires a certain amount of skill. Competition among vendors for space in the streets and access to customers is strong in many cities. And vendors must be able to negotiate effectively with wholesalers and customers.

Street trade can offer a viable livelihood, but earnings are low and risks are high for many vendors, especially those who sell fresh fruits and vegetables. Having an insecure place of work is a significant problem for those who work in the streets. Lack of storage, theft or damage to stock are common issues.

By-laws governing street trade can be confusing and licenses hard to get, leaving many street vendors vulnerable to harassment, confiscations and evictions. The IEMS research found that even vendors with a license had trouble finding a secure vending location, and those following the regulations sometimes had their goods confiscated. Learn more about Street Vendors and The Law.

Occupational Health and Safety

Working outside, street vendors and their goods are exposed to strong sun, heavy rains and extreme heat or cold. Unless they work in markets, most don't have shelter or running water and toilets near their workplace. Inadequate access to clean water is a major concern of prepared food vendors.

Street vendors face other routine occupational hazards. Many lift and haul heavy loads of goods to and from their point of sale. Market vendors are exposed to physical risk due to a lack of proper fire safety equipment, and street vendors are exposed to injury from the improper regulation of traffic in commercial areas.

Insufficient waste removal and sanitation services result in unhygienic market conditions and undermine vendors sales as well as their health, and that of their customers.

Vulnerability to Economic Downturns

Economic downturns have a big impact on vendors earnings. In 2009, an Inclusive Cities research project found many street vendors reported a drop in consumer demand and an increase in competition as the newly unemployed turned to vending for income.

A second round of research, done in 2010, found demand had not recovered for most vendors, and many had to raise prices due to the higher cost of goods. Competition had increased further as large retailers aggressively tried to attract customers.

The 2012 Informal Economy Monitoring Study confirmed that rising prices and increased competition were still affecting street vendors in several cities. Vendors said their stock was more expensive, but they had difficulty passing on rising costs to consumers, who expect to negotiate low prices on the streets. More competition means vendors take home lower earnings.

What is Vending Management?

Vending management is a single-source solution to providing vending services for businesses with multiple locations and consolidating them into one account. A vending management company uses its network of preferred vending companies like Coca-Cola, Pepsi, or local full-line operators to install and maintain vending machines at each property.

This centralized approach keeps your organization focused on core business responsibilities while still benefiting from the convenience of having vending for all locations. All the service and repair issues are managed by one company who knows your account well and can easily provide reporting when needed.

Without a vending management program, your organization has to work with many different vending companies because they all operate on a regional basis. Even Coca-Cola and Pepsi, despite their global presence, are divided among territories and different ownership. As a result, they do not have the infrastructure to manage a national vending plan. This fragmented approach can cost your employees time away from what they really need to focus on. Vending management eliminates that hassle by providing one point of contact for all vending machine requests.

Loose change adds up. A vending management program takes the commissions from each vending machine at each property and consolidates it into one check. This relieves personnel from having to sift through dozens or hundreds of commission statements. Plus, a vending management program will keep track of commissions. So if some are missing, the vending management company will be sure to collect them. What sounds better: a bunch of small checks scattered about your organization, or one big check sent quarterly?

Transportable

First, you have to know some of the pros of street vending business. The first advantage of this kind of venture is that you will have the chance to evaluate the area you are planning to establish your business. Since vending business can be transported from one place to another, you can easily relocate to the other places once you seem that the current market you have will not offer you with the profits you are targeting. This is not like the other complex businesses where the location they currently have will be fixed.

Low Overhead

Aside from being transportable, street vending business will also offer low overhead for its business owners. In this kind of business, the vendors do not longer need to secure payment in electricity, business rentals and other financial responsibilities of the other kinds of businesses. This also follows that it will easily lead to the generation of higher sales. The income that will be earned does not longer need to be allotted with the other business financial responsibilities.

Unlike the other kinds of business, there is a need to pay for electricity bills, location rental and other aspects.

Neck to Neck Competition

One of the main disadvantages in street vending business is its neck to neck competition. Once you established this business, there are so many street vendors that may come. This may be the reason why these kinds of businesses are made transportable because they need to go to one place to another in order to look for the prospects of their business. And because of the competition of this kind of business, there is difficulty in searching for the right business location.

Exposed to Harmful Weather Conditions

On the other note, the street vending business owners are very much exposed to health alteration. This is a form of venture where you are frequently exposed outside. You will suffer much from the extreme weather conditions like excessive rays from the sun as well as rain. If you are a vending business owner and have altered immune system, you will easily acquire diseases and infections such as colds and cough. But in order to avoid this, you can take some medications that will boost your immune system response.

Best Street Vending Business ideas & Opportunities

Start an Ice Vending Machine Business

Ice vending is one of the street vending machine related business ideas that can thrive in the United States of America. An ice vending machine business is a retail business that does not require face to face interaction with customers, and can be run for 24 hours a day depending on the location.

This business offers its operators flexibility, because it can be tailored to suit the lifestyle of the owner. Anyone can start an ice vending machine business; it isn't financially tasking to start, as you can purchase a used ice vending machine and still get the most out of it. The ice vending machine business doesn't require any expertise or know-how, and any serious minded entrepreneur is likely to make good profit out of this business.

Groceries Vending Machine Business
Groceries vending is one of the street vending machine ideas that can be started in any country of the world. Starting a grocery vending machine business is a very easy business to start and it is not so capital intensive. Getting the right products that people want to buy is the secrets of running a grocery vending machine. In order to maximize profits from this business, just ensure that you position your vending machines in strategic places with good human traffic.

Drinks Vending Machine Business

One of the most popular street vending machine businesses in the United States is drinks vending machine; a vending machine that dispenses chilled bottles of soft drinks and energy drinks.

You can find this type of vending machine in stadiums, public facilities, schools, office lobbies and hotel lobbies amongst others. So, if you are looking towards starting a vending machine business, then you should consider starting drinks vending machine business.

Snacks Vending Machine Business

Another popular and no doubt lucrative street vending machine related business that can be started in any location is snacks vending machine business. A snack vending machine is a vending machine that dispense freshly backed snacks.

If your snack vending machine is loaded with snack foods such as potato and corn chips, pretzels, roasted and salted nuts, nut butters, popcorn and other similar snacks, you won't struggle to generate sales. The bottom line is that a snacks vending machine business is bound to thrive if properly positioned.

Children's Toys Vending Machine Business

Starting a children's toys vending machine business is yet another good vending machine business that can be started in any country of the world. A children's toys vending machine is a machine that dispenses various children's toys. In as much as this is a profitable business, but note that children's toys vending machine can only thrive in strategic locations; places like children's parks, schools, playgrounds et al are suitable for this kind of business.

Beauty Products and Cosmetics Vending Machine Business

Another lucrative street vending machine business that an aspiring entrepreneur who is interested in starting a street vending machine related business should consider starting is a beauty products and cosmetics vending machine business.

If you are considering starting a beauty products and cosmetics vending machine business, then you must ensure that your vending machine is stocked with a wide range of hair care and shower products, cosmetics, skin care products, fragrances, nail care products, deodorant and shaving products, sun care, baby care and other related products et al from different manufacturers both from the United States of America and from other countries. That's only when you will continue to welcome repeated customers and of course grow your profit margins.

Start a Fruits Vending Machine Business

Fruits vending machine business is another flourishing street vending machine related business that can be started in any country of the world. Fruits vending machine is a vending machine that dispenses fruits. In order to maximize profits from this line of business, aside from ensuring that your vending machine is well-located, you also need to ensure that your fruits are fresh and if possible organic.

Start a Vegetables Vending Machine Business

Similar to starting fruits vending machine business is starting vegetables vending machine business. Vegetable vending machine business is a viable and lucrative street vending machine business that can thrive in most locations. In order to maximize profits from your vegetable vending machine business, then you must ensure that your vending machine is stocked with fresh vegetables.

Start a Handkerchiefs and Towels Vending Machine Business

Another simple but of course lucrative street vending machine business that an aspiring entrepreneur can start and make money from is handkerchiefs and towels vending. In the United States, this type of business does pretty well during summer.

This goes to show that if you are planning on launching handkerchiefs and towels vending machine business in the United States, then you should expect to make more sales during the summer. You can as position your vending machines in places like gyms, and sports centers; places where people are bound to sweat.

Start a Flowers Vending Machine Business

A flower vending machine business is yet another profitable street vending machine business that an aspiring entrepreneur can make money from. A flower vending machine is a vending machine that dispenses fresh and artificial flowers. So, if you are looking towards starting a street vending machine business, then you can settle for flower vending machine.

Lottery and Events Tickets Vending Machine Business

Another easy to set up street vending related business that can be started in the United States of America is lottery and events tickets vending machine business. The fact that most people would rather wait for late hours to purchase their event tickets makes lottery and events tickets vending machine business a viable business to run. In order to maximize profits in this line of business, then you must ensure that you partner with key events managers/organizers and lottery companies so as to help them sell their tickets from your vending machine.

Alcohol Drinks Vending Machine Business

Alcoholic drinks vending machine is one of the most popular street vending machine business you can find around. A standard alcoholic drinks vending machine is expected to have assorted beers, wines, liquors, distilled spirits, and martinis amongst others.

Please note that there are very strict laws regarding the sale of alcohol through vending machines due to concerns regarding underage buyers. Just ensure that you position your alcoholic vending machines in night clubs, and any other locations where adults relax and you won't struggle to generate sales.

Cigarettes Vending Machine Business

Having a cigarettes vending machine business is another lucrative street vending machine business that an aspiring entrepreneur can successfully start. If you are planning on starting this type of business in the United States of America, you have to check the existing law so as not to run afoul of the law.

This is because there are very strict laws regarding the sale of cigarettes through vending machines due to concerns regarding underage buyers. In the uk for instance; cigarette vending machines have been banned, and in Japan, Germany and Italy, age verification has been made mandatory.

Music and Movies Vending Machine Business

There is indeed a large market for music and movies and this market cuts across people of different faith, race and culture. The United States of America has the largest entertainment industry in the world and products from this industry can be retailed via vending machines.

If you live in the United States or in any part of the world and you are considering starting a street vending machine business, one of the businesses that you can successfully start is music and movie vending machine business. In other to stay competitive in this industry, you must vend music and movies in all formats; that is in softcopies (downloads) and hardcopies (CDs and DVDs et al).

Cupcake Vending Machine Business

Another street vending machine business that an aspiring entrepreneur can start is cupcake vending machine business. It is a known fact that some cupcake bakers have started retailing their cupcakes to the public through vending machines just like the owner of Sprinkle Cupcakes, Candace Nelson did. In order to maximize profits from your cupcake vending machine business, you must ensure that you have cupcakes in different designs, themes, and flavors and of course different packaging in your vending machine per time.

Frozen Food Vending Machine

Frozen food vending machine is another typical street vending machine business and it is indeed a thriving and profitable business venture. Frozen foods such as fish, chicken, turkeys et al are generally consumed worldwide. So if you are looking for a street vending machine business to start; a business that is not capital intensive and simple to manage, then you need to consider setting up a frozen food vending machine business.

It is important to state that the location you choose to start this type of business will go a long way to determine if the business will become successful. The ideal places to locate this type of business are market places or residential estates et al.

Bread Vending Machine Business

A bread vending machine business is another lucrative street vending machine business that an aspiring entrepreneur should consider starting. Bread is a stable food that people eat in all parts of the world. If you are looking for a street vending machine business to launch, a business that requires little capital to get started, then you should consider retailing bread via vending machines.

All you need to do is to secure deliveries of bread from different bakeries and then display them in your well located vending machines. It is a cool way of making money but you must ensure that you sell them all as fast as you can because breads have short shelf life.

Ice Cream Vending Machine Business

An ice cream vending machine business is another cool street vending machine business that any serious minded entrepreneur can successfully start and make cool cash from. It is a business that requires little training and skills to set up and manage.

If you are looking towards starting this type of business and you don't have the skills, you can quickly learn them, it is likely not going to take you more than a week to learn how to make different flavors of ice cream. Even if you don't have the required capital to purchase brand new vending machines, you can start with fairly used vending machines.

Hair Extensions and Hair care Products Vending Machine Business

It is observed that ladies spend more on shopping when compared to their male counterparts which is why any business that is geared towards ladies would thrive. If you are looking for a street vending machine business to start, a business that is easy to start and manage, then you should consider vending hair extensions and a wide range of hair care products. It is a thriving business and profitable business. Just ensure that you choose a good location for the business, places like salons and female hostels is ideal for such vending machine business.

Bottled Water Vending Machine Business

One of the most popular street vending machine business in the United States is bottled water vending machine. You can find this type of vending machine in stadiums, public facilities, schools, offices and hotel lobbies amongst others. So, if you are looking towards starting a street vending machine business, then you should consider starting bottled water vending machine business.

Chocolate Bar Vending Machine Business
Another business that vends on street corners that entrepreneurs could think about beginning, a business that is simple to start with minimal initial capital investment could be the chocolate-bar vending machines. This kind of enterprise is very easy to run and it's certainly a lucrative venture, especially in the event that you conduct detailed studies of feasibility prior to choosing the location for the vending machine. There is no need for special abilities to manage this kind of business.

Children's Educational Materials Vending Machine Business

vending machine that distributes educational materials for children is another business that street vending machines run that entrepreneurs can start with no technical knowledge. Materials such as types of books, learning pads, stories, Lego, jig saw puzzles, educational video games, educational CDs and DVDs etc are very lucrative as they are placed in a company in an area where children are in the vicinity.

You could also place vending machines inside schools. Selling educational materials to children is a lucrative business that anyone can get into this kind of business. In fact you don't require any permit or license to operate this kind of business.

Perfume Vending Machine Business

The selling of perfumes by various designers is another profitable vending machine on the street which entrepreneurs can think about to start. This kind of business is straightforward to begin and is extremely easy to manage. The most important thing you have to complete to get the business up and running is to acquire vending machines, places for the vending machines and fill the vending machines with a variety of fragrances designed for the market. Like the majority of street vending machine companies it is important to choose the right location. If you're able to locate your machine in a suitable location, perhaps in a bustling lobby it will be easy to generate daily sales.

Frozen Yogurt Vending Machine Business

The business of vending machines for frozen yogurt is yet another profitable street vending machine-related business that is available to anyone and everyone. Yogurt is consumed by virtually everyone on the planet. As therefore, anyone who decides to set up an ice cream vending machine anywhere in the world of the globe, particularly in countries with tropical climates or located near the equator are will reap good profits from their investment if they do exactly what they should. If you're thinking about starting a vending machine company, one option is to establish the frozen yogurt vending machine company.

French Fries Vending Machine Business

A French vending machine that sells fries is another popular street vending machine business opportunity that aspiring entrepreneurs is able to start and earn decent money from. In a ideal

French vending machine for fries, you will find products like potato chips, corn chips, pretzels salted and roasted nuts popcorn and other similar snacks.

They are the most popular street food that are eaten by millions of people, not only in France but across the world. If you're contemplating starting a street vending machine-related company, then look into beginning the French vending machine for fries.

Dessert Bar Vending Machine Business

The idea of starting an automated dessert vending is another viable street vending machine company that any aspiring entrepreneur could be successful with the help of the United States of America. Dessert bars, also known as bars or squares are a form of American cookie with the firm texture of a cake, or is softer than the typical cookies.

They are cooked in a pan before being cooked in an oven. The main ingredients to make desserts bars include eggs, sugar as well as butter, flour, milk , among others. The more exotic bars are made using ingredients such as Rhubarb, sour

cream pretzels, candy, raisins, vanilla and
pumpkin.

Biscuit and Candies Vending Machine Business

Another street vending machine business which
aspiring entrepreneurs with a small amount of
capital to be able to start is a the vending
machine for biscuits and candies. If you're
thinking of setting up a biscuit and candy vending
machine, it is best to locate the business near an
elementary school or children's playground.
Children are the primary buyers of candies and
biscuits. Be sure the vending machines are filled
with a variety of candy and biscuits, as well as
drinks for children and you'll never have to
struggle to make sales every day.

Start a Photo Booth Business

The idea of starting a business for a photo booth
is definitely one of the lucrative and well-
established street vending machine business
ideas that entrepreneurs who are looking to start
is able to successfully launch and earn profits
from. When we speak of photograph booths
we're discussing a vending machine , or a modern
kiosk with an automated, typically operated by a
coin, camera and processor. A significant portion

of the photo booths currently operational across the United States are digital. If you're thinking about beginning a vending machine-based business then one option is to establish the business of a photo booth.

Eggs Vending Machine Business

Another possible street vending machine service that entrepreneurs who are aspiring can establish with in the United States of America is egg vending machine. A vending machine in which fresh or boiled eggs can be offered will certainly succeed if placed. If you're searching for a street vending machine to begin a business with, you must think about beginning a business with eggs.

Coffee Vending Machine Business

If you're considering the possibility of starting a vending machine on the street business one of the options that you have is to establish an enterprise that involves coffee vending machines. The process of starting a coffee vending machine business is interesting and at as well as rewarding when you place your vending machines for coffee in strategically placed places that will draw in coffee enthusiasts. The fact that many people drink coffee at the go is one of the reasons why it is among the businesses that is highly sought-after.

Birth Control (Condom and Contraceptives) Vending Machine Business

Birth contraception (condom as well as contraceptives) vending machine company is another lucrative street vending business opportunity that aspiring entrepreneurs could start and earn money from. Birth control machines are an automated vending machine used that sells contraceptives for birth control, like condoms as well as emergency contraception. Machines for vending condoms usually located in hotels, brothels stripper clubs, campuses and public toilets. They are also found in airports, subway stations and schools as a precautionary initiative to encourage safe sexual sex. Some pharmacies also have one outside for access after hours. There are a few instances of pharmacies that offer condoms for females, or even an after-dinner pill. Adults looking to stay clear of sexually transmitted illnesses are advised to use condoms. This is a reason to sell condoms.

Dietary Additives Vending Machine Business

A distinctive street vending machine service which aspiring entrepreneurs ought to look into is vending machine for food supplements. A vending machine for food supplements is a device that offers supplements, such as vitamin and mineral capsules or tablets special nutritional supplements, herbs and plants, protein powder meal replacements, weight loss products, special and elite sports nutritionals , et and so on. It is important to research the current law regarding prescribing medicines without prescriptions prior to launching your vending machine for food supplements business.

Stationery Vending Machine Business

Stationary vending machine companies are an additional street vending machine company that is able to be established in any part of the world. If you are planning to maximize profits of your stationary vending machines enterprise it is essential to be willing to put the machines in exam centers or job centers, as well as offices and schools.

You must keep your vending machines stocked with stationery supplies, such as pencils (black blue, red and green) erasers, permanent markers sharpeners, pencils as well as colored pencils, colored markers, colored pens and correction tape/fluid/liquid papers mechanical pencils and spare leads in addition to.

The Change Vending Machine Business

Another well-known street vending machine company that investors can begin with in the United States is change vending machine, or better change machine. The term "change machine" refers to a machine that takes large denominations of currency and provides the same amount of currency in smaller denominations, such as coins or bills.

They serve to offer cash in trade for currency in paper, which is the these machines are called bill changers. If you're planning to start an enterprise that involves vending machines one of your choices is to begin a the business of vending machines that change.

Start a Full Line Vending Machine Business

A full line vending machine company is another extremely profitable business of street vending machines that is able to be set up in any part of the world. A full-line vending firm can create a variety of vending machines which sell many different products. These can include cookies, candy chips, chips, fresh fruits milk, cold foods coffee, liquids like hot beverages and bottles and cans of soda, as well as frozen goods such as Ice cream.

Newspapers and Magazines Vending Machine Business

A more lucrative street vending machine company that aspiring entrepreneurs should think about is the newspaper and magazine vending machine businesses. A newspaper or magazine vending machine machine designed to dispensate magazines and newspapers. Vending machines for magazines and newspapers are widely used across the globeand could be among the principal distribution methods for magazines and newspapers.

and Train Tickets Vending Machine Business Bus and Train Tickets Vending Machine Business

The idea of starting a train and bus ticket vending machine is among the most mobile street vending machine business concepts that can be launched within the United States of America. Tickets vending machines are machine that makes tickets.

Tickets are sold by ticket machines at stations for trains as well as Metro stations sell transit tickets as well as tram ticket tickets in a few tram stops as well as in certain trams. The normal transaction is users using the interface for display to choose the type or quantity of tickets, selecting a payment option that includes cash or credit/debit cards or smartcards. Tickets or tickets are printed and handed out to the person who purchased them.

Stamp Vending Machine Business

Stamp vending machines are another street vending machine company that is thriving in all nations around the world. A stamp vending machine an electrical, mechanical or electro-mechanical device that is used to offer postage stamps to customers in exchange for a specified amount of money, usually in the form of coins. If you're thinking of starting your own street vending machine company it is best to opt on a stamp vending machine company.

Pizza Vending Machine Business

A fast-growing vending machine company that is gaining traction within America United States of America is pizza vending. Let's Pizza is the first vending machine that produces homemade pizzas from scratch. It was created in 2009 by the Italian Sitos srl, a company from Italy.

The machine blends flour, water as well as tomato sauce and other fresh ingredients to create the pizza in about three minutes. The machine has windows so that customers can view the pizza being created. The invention was made in the hands of Claudio Torghele, an entrepreneur in Rovereto, Italy. You could start something similar.

Books Vending Machine Business

The business of vending books is another vending machine on the street which is growing in popularity in the developed nations around the globe. Book vending machines dispense books. Certain library systems in the United States make use of vending machines for books. GoLibrary is an automated book vending machine that is used at libraries across Sweden as well as the U.S. Therefore, if you're seeking an opportunity to vend machines for your business to begin, you should think about settling on vending of books.

Bait Vending Machine Business

A different street vending machine service that entrepreneurs who are considering to start is a bait vending company. The bait vending machine can be described as a machine that offers baits for fishing that are live, like crickets and worms for fishing. If you wish to make the most profits from your business, make sure that you place your vending machines close proximity to fishing spots.

Marijuana Vending Machine Business

An marijuana vending machine enterprise is yet another lucrative street vending machine enterprise which can be legalized legally in the United States of America especially because of the new legislation that supports the legalization of marijuana for recreational use in a few states of the U.S.

The marijuana vending machine can be described as a machine that sells or distributes cannabis. In the decade 2010s they were in use throughout Canada and the United States and Canada. The main challenge in selling controlled or restricted products like cannabis is proving that the buyer purchasing the product, which is solved by the use of biometrics and smart vending technology.

Social Networked Vending Machine Business

The social networking vending machines business is a brand new street vending machine company which can be launched in countries that are developed in the world. With the advent of social media vending machines are linked to social media to increase the number of interactions between the vending machine and the people who use the physical machine on social networks. The most common use of a social network-connected vending machines is that the customer is able to connect their social accounts to a particular social network that is identified to the machine. the customer will get certain rewards usually in the form of a free gifts that are dispensed by the machine.

Cards Vending Machine Business Cards Vending Machine Business

The vending machine for greeting cards is another lucrative street vending machine company which aspiring entrepreneurs must think about setting up. People love cards. They are enthralled by the warmth and love that emanates from the messages communicated through the cards.

Another thing that catches the attention of people with cards is the combination of colors and the images that are captured. To make the most profit in this field of business, it is essential that you have stocked your vending machine numerous greeting cards like birthday cards, get well cards, congratulations cards, thank you cards baby shower cards, condolence and consolation cards, and more.

Installation and Supply of Vending Machines

If you're thinking of beginning a vending machine related business, one of the choices, particularly if you're not keen on selling through vending machines , is to establish an organization that focuses on the installation and supply of vending machine. I guarantee you that there are a lot of retailers who will willingly purchase vending machine from you. It's trending for retailers to use vending machines to boost sales.

46. Repair, Maintenance, and Servicing of Vending Machines

Another arduous and lucrative vending machine-related enterprise that entrepreneurs who are considering getting into is the repair, maintenance and upkeep of machines. In reality, vending machines suffer from wear and tear, and when they become malfunctioning they must be repaired.

If you're seeking an enterprise that deals with vending machines to start , and you're aware that you have an engineering experience You can begin the vending machine repair maintenance, and maintenance business.

Easy Steps to Become an authorised Food Vendor
Be aware of the law that governs Food Vendors
The first step in order to become an authorized food seller is understand the current laws governing food sellers operating in the city where you are planning to operate, in the state where you are planning to operate and all across the United States of America. It's advantageous to be informed prior to setting into obtaining an official license for your business.

The most important step to complete to receive the information you need about the law that regulates food sellers across the United States of America is to go to the Department of Consumer and Regulatory Affairs, Vending Division. They are more than willing to help you.

As an example for instance, for instance, the Department of Consumer and Regulatory Affairs is not currently authorized for the issue of street vendors licenses to those who plan to operate on sidewalks in cities. Therefore, if this is the type of business you're planning to start and you want to start, you must reconsider your strategy or you will not have the necessary license for food vendors.

Start a Business

If you think about it No one can give you a license to run an enterprise that isn't registered. If you've done your research and are prepared to begin and run your own food company within the United States, then you have to start by registering your company.

Be aware that if your business is an entity that is a corporation, Limited Liability Company, partnership, or in certain instances, a trust, the organization you are forming is required to be registered and at a good standing in the Corporations Division. For more details, including assistance with the application it is required to call for assistance with the Corporations Affairs Division in your state.

Secure a Kitchen Facility

The next step in being a registered food service to the United States of America is to locate a kitchen. This is crucial and is a good idea to do it in conjunction with the registration of your business since you'll need an address in order to register your company and, of course, in the event of an inspection by the department responsible for regulation.

In actual fact, every applicant seeking a Basic business License for the category of Street Vendor A category need to be evaluated by and/or get approval by the Department of Health, Food Safety & Hygiene Inspection Services Division and the Department of Consumer and Regulatory Affairs, Vending Division.

You must construct your kitchen to conform to the standards of food safety, health and hygiene before you are able to be licensed as a food seller in the United States.

Request an Food Vendor License

If your kitchen facilities have been examined and you've been granted approval by the regulator this means you are in good shape and you are able in applying to get a license as a food vendor.

The essential requirements to be able to provide prior to applying for a license as a food seller include a business registration certificate and a kitchen facility, mobile food truck, or kiosk, an the application fee and form.

It is vital to mention that each state within the US has its own set of requirements regarding the issuance of food licenses to those who apply. For instance In Washington DC, the license expires after for two (2) two (2) years (you must renew your license as a food vendor at every 2 years). The license is not accessible online. You will have to visit their office to apply for and get the license. If you're looking at the fee, here's how it is (Category Cost of License: $338.00, Application Fee: $70.00, Endorsement Fee: $25.00, Technology Fee: $43.30 and the Total Cost: $476.30).

If you satisfy the standards as described above, you'll be able to obtain your food vendor's license and then become a licensed food seller across the United States of America. It is essential to note that certain states have less strict requirements for getting an licensed food vendoraEUR(tm)s license. That means that if you do not meet the required requirements to be licensed food vendor in one state or city then you may try different cities or states.

Chapter 2: An Effective Business Model In A

Vending Business

The next chapter we'll be discussing a profitable business model for vending. What exactly does that mean? In the past few years there has been a dramatic rise in the number of food and beverage companies within North America. This is typically because of technological advancements and the introduction of new products into the market. One thing that is rarely considered when expanding the scope of a company is the method of making a business plan that is effective.

To develop a viable business model for your vending company, you should first examine the company's goals and goals and take into account the external factors that can affect the company's success or demise. If you are looking to establish a profitable business which does not require external funding, it is prudent to look into the Micro-Loan Program offered by the SBA.

A variety of conditions must be met for a business to be eligible for a micro-loan through this program. For instance, the firm must be a small business with less than 25 employees, but must not more than 100 employees. The business must show an annual net profit that is steady and have a five-year financial record. The business must also be owned by a private company and not a subsidiary of a larger company. There are additional requirements the company must satisfy to be eligible to be eligible for loans under this program, for example having a solid business plan in place as well as being able to show that it is able to produce enough earnings to cover the operating costs, not having tax owing, and presenting the evidence of a credit history that is good.

When you are putting together a successful business plan for your professional vending business, you must be mindful of the three I's of innovation Integrity, Integration and Interaction. For a successful business model, your company must provide value to your customer and have a competitive edge over competitors in your field. The three steps listed above are vital to develop a viable as well as profitable model for business in any sector.

To provide worth to customers it is first necessary to know what your customer desires or requires. It may sound to be common sense, but often businesses overlook that they're selling products and services, not because they're selling products or services, but also because they're creating the best standard of living for their clients. In order to create a profitable business, you must to make sure that your company will help people live happier lives. This could mean aiding them in saving money or increasing their profits. It could also be about giving their customers a better quality life in terms of convenience and accessibility of services and products.

Monitoring customer preferences is possible through market research. One of the best methods to gather details from your customers on preferences is to conduct focus groups. Focus groups allow participants to ask questions about various products and services for your company to be more efficient in providing the information that your customers desire or require. For focus groups to be successful it is essential to provide an equal opportunity for everyone in the group. For example, if you were to organize a focus group about the topic of sleepwear and mattresses, it should not be able to exclude males from participating in the discussion.

Another way in which your business can provide value to customers is to incorporate new products in your portfolio that are in line with the preferences of your customers. The most effective method of doing this is first figuring out who the customer is and then determining what they require or desire. If you've already conducted market research it will be much easier to understand the needs of your customers since the research will reveal what they prefer in terms of goods and services. Another way to research the preferences of customers is to use survey responses from company employees.

In order to provide more value to your customers you must also know the preferences of your customers when it comes to the manner in which you run your business. This means that you take the time to hear your clients and then responding in a manner that is appropriate. If you can provide better quality customer services, you will gain advantages over competitors that are in your industry. To provide high-quality services and products at a reasonable cost, you require an efficient business plan that can allow for greater production and efficient operations.

In the process of integrating and introduction of new items to your company All aspects of production have to be considered. For instance, if you're planning to introduce new products into your company You must take into consideration all the aspects involved in the production of these products. This includes the price for raw material, the price of labor and energy to make the product and the processing costs.

A well-functioning supply chain is a second strategy for your business to deliver better value to your customers. It's not uncommon to find numerous organizations to be involved in production of raw materials, from sellers and distribution companies to retail stores and customers. One method for companies to provide higher-quality product at a lower cost is to create efficient chains of supply that permits companies to manage their labor and transport costs.

When you are developing a business plan, your client should be aware that it's not just about the services or products you provide , but also the way you run your business. In order for a customer to get the most value from your business both parties must be in agreement. That is to say, the company has to be committed to satisfying the requirements of its customers, while customers have to commit to purchasing products from the business. If a business meets these obligations, they will be satisfied customers and they will be able to meet their goals of making a profit from their business model. To allow this processional vending business to be successful, it is essential to incorporate all of these aspects when creating a profitable business plan.

In establishing a successful business model for a successful vending business, it's essential to identify the goals that must be met. In order for a model to be successful, it should achieve certain milestones, like profits margins and growth in revenue. Another aspect that is crucial to creating an effective business plan is understanding the company's advantages and disadvantages. In addition, you need to establish internal measures of success that are based on the achievement of goals and external measures of success that are based on market research regarding preference and share of market.

In the end, you need to create your business plan that takes into account all the possible revenue streams for your business. This will enable your company to identify future expenditure and revenue trends to respond according to the trends. Furthermore, the best method of analyzing every revenue stream is by using an official pro model. Pro formas allow you to evaluate whether your model of business is sustainable and identify areas you'll need to make improvements to make sure that every aspect of your company are being considered.

Chapter 3: The Price And The Location Of

Considerations

Vending machines are an excellent idea for a business if you're hoping to earn a profit without investing any effort or time. In general, vending machines is able to be located anywhere, from airports and malls to just down the street. With the variety of vending machines available it could be difficult to determine which is right for you. In this article I'll go over the factors that affect pricing and location and offer suggestions for selecting the right machine to meet your needs.

Good Placement for Vending Machines

In assessing the location of vending machines when deciding on the location of vending machines, it's important to consider the kind of location you choose and the type of patrons who come to the place. For instance the mall is the best place to sell drinks or snacks because these are the things that people typically are looking for when shopping in the mall, and they'll likely be able to purchase in the process. In certain situations you may find it more economical to pay for your space inside an area of shopping instead of renting from another.

It is possible to put several machines within a single business in various places. If, for instance, you have a retail shop and a barbershop, side by the other, you'll require a machine for both shops. In this scenario you may choose to utilize only one vending machine for the barbershop, and another set of machines for the store for retail.

The location is also crucial in the setting up of several vending machines in the town since you don't want find one close to an elementary school or in an area where it could be vandalized. Look for places which are ideal for the location of your sales zone and ensure you're not placing machines in areas with busy roads or areas of residential in which people could take or destroy your equipment.

Pricing Factors

When determining the price of the vending machines, you have to consider numerous factors that can influence your profits. Here are a few important things to consider:

* Location. It may not be required to place a machine located in an office corner, however it's never worth it to purchase vending machines if it does not sell well. If you buy one that doesn't perform well, you may have repair it and then utilize the time to use it for other things.

* Customer counts. A good rule of thumb is to set your prices at least two-to-three times more when you're in cities than on the upper reaches of the city. The reason is that if there are lots of customers in a certain region, you'll be able to earn more profit for each machine than when have less customers.

* Compliance with State Laws. Each state has its own laws that govern vending machines. You should be aware of the laws and how they apply to you. If the vending machine isn't selling something, certain states may be able to consider it a vending machine, which could result in fines and other penalties.

* The size of the machine. If you're thinking about purchasing vending machines the dimensions that the device is going to impact the amount of money you can earn.

The sleek and compact machines are great for public spaces that have limited space however, they're not as lucrative as larger models. If your business is big enough that it warrants a huge vending machine (both in terms of quantity and profits) then by all means pick one that is and is ready to be put in the most prominent place. Keep in mind that this decision could cause higher costs for each machine that is purchased.

Location considerations

There's nothing more crucial than making it perfect the first time around when placing the machine. It's not a good idea to buy an item that isn't selling or is vandalized and so on. If you're unable to spot mistakes when you're putting in the machines in your home, then you could be tempted to throw it away and purchase a new one. To assist you in the decision-making process Here are some suggestions which could help:

* Stores. If you're planning on purchasing an retail storefront that is close to machines for vending, you'll have be aware of the type of customer who will frequent your establishment. For instance, if your location is mostly an establishment for barbershops, it's probably be better to get an ice

cream machine then an iced coffee machine or coffee machine since those kinds of drinks will be sold more frequently in these kinds of shops.

* Hospitals. If a hospital is located close to the vending machine you use it is possible to think about purchasing health-conscious snacks and drinks such as Power bars and bottled water.

* Colleges. If a college is located close to where you live You might want to think of snacks that students might want, such as candy bars, chips and so on.

In some cases, the location can be identified by the dimensions of the area or street you're looking at and also where it is situated in relation to other areas of residential or commercial activity. Look for places with good lighting at nightand have easy accessibility from parking areas and not located directly on busy streets or along main roads.

If you are considering the location for vending machines take into consideration the factors of space and traffic flow. You should also consider how easy customers will be able to travel to your location. In the majority of cases it is best to be in close proximity to other businesses, so that customers will be able to come quickly.

Chapter 4: The Location In A Vending Business

In the world of vending it is the location that counts. To be successful, a business must have a good location there must be an adequate number of customers who could be at the correct location at the appropriate moment. If a vending business locates its machines in the wrong place and fails to generate enough profits to be profitable. Many people do not know how crucial it is to have the machines that generate income (IGMs) to sit located in the best places. It could determine their profits and performance when they try to locate new sites that are ideal for the flow of traffic and foot patterns of customers who may be interested in purchasing drinks or snacks from their IGMs.

If you're an enterprise that sells vending machines is crucial to understand how you can optimize your facilities.

The most crucial location-related elements are:
* Foot Traffic If a vending device is located near the entrance of an establishment or close to an elevator, it'll receive lots of pedestrians. If it's located in a remote location without any foot traffic and no foot traffic, it will result in lower sales , and consequently less profit. Make sure that IGMs are situated in areas that pedestrians pass across the street all day long. When selecting an area be sure to look for areas that are frequented by people like entryways, escalators, and elevators.
* Number of passersby The sales volume generated by IGMs rises with the number of people passing by. For instance, if there are two vending machines located in the same place, however one one has significantly higher visitors than the other the machine will make double the amount of sales as the other machine since it will have twice as many people passing by. Therefore, you should look for places that have a high volume of traffic and lots of pedestrians.
The Crowd Density - As the rule of thumb you should have large crowds at vending machine locations in order to guarantee that they can

generate decent sales. Crowd density is the measure of how tightly packed the crowd is within an area. More tightly packed crowds result in greater numbers of people walking by and a higher volume of sales from IGMs in these locations.

* Store Competition If there's no vending machine competition You'll be monopolizing the market and therefore be able charge higher rates. If there are other machines operating in the vicinity the profits and sales will be less. Also, ensure that there are other IGMs nearby so that your profits aren't eroded due to competition.

* Near Food and Drink Outlets The best places for IGMs are in close proximity to restaurants or beverage outlets since there is always a need for something whenever they're in public. If they're short on time or don't have food available or food, they'll go to vending machines. Be on the lookout for eateries close to your IGMs so you can increase sales and foot traffic.

* Urban or Suburban Location A different type of urban area is a suburban. Urban areas are ideal for vending machines as they're conveniently located and attract plenty of foot traffic,

especially from pedestrians. Suburban areas are usually not aware that there are IGMs. As a result, they're less likely to be being visited by those who are in need of snacks. Therefore, keep in mind areas that are rural or suburban in areas with the possibility that potential clients will purchase drinks and snacks from vending machines in close proximity to them.

Websites such as 0rchard Vending can take care of the heavy lifting for you. They provide the most effective online portals to find locations and placing IGMs in busy zones.

What happens when you live in a place that makes everything Different

The location is among the most crucial aspects of vending. It is a factor that determines the profitability of your business, sales, customer base as well as other. When you're entering an unfamiliar market in the beginning, it may be difficult to determine what you need to do.

New markets have a high potential for profit however they need a lot of effort. What happens if you attempt to sell the same products with a new place? Should you combine the numbers from your previous location and the new one? Should you take the average from both locations? This book will show you how to select the most suitable location to sell your products. After you have read this book you'll know how to evaluate two locations with similar functions.

Every company has to determine if it is in the right area or another. If it doesn't, it will not exist! What are the methods managers can select the most suitable areas for their business? Let's talk about it.

Location Selection

There are a variety of factors that determine what a business's best option is to locate in one area or in another. Let's look at them in depth.

* Profitability: Profit is the top priority for the location of a business. The term "location" refers to the time the place where goods are kept in a warehouse, sold, or shipped to clients. Vending is classified as a service-based industry since the business provides products and services (profit). If a shop isn't profitable and doesn't earn money, it's not profitable and if you can't earn money, you will not last for long.

* Market Penetration Market penetration is the amount of people that the product you offer is able to get to. When market penetration rises the profitability will also increase. That means that if your pie is large enough, you are able to sell more hamburgers for less cost because the number of customers who are able to purchase the hamburgers is increasing. Therefore, the greater number of people you are able to reach within your market you'll be more successful in your business will be.

• Market analysis: The analysis of the market is crucial prior to making a decision about "which location to enter." It's essential before deciding which one to choose and should be carried out every time one plans to open a new store. It is about knowing the potential market and selecting the best location in accordance with which has the highest potential for sales compared to the other.

Location Factors

Many factors influence the best location for your vending business.

"* Average Sales." This is the average amount of units that you sell within an hour.

* Machines Number: The amount of machines you own makes an enormous difference to your profits. If you have more equipment own and the higher the profits you'll earn since you'll have the ability to introduce more items to market with low prices.

• Setup costs and rental Setup costs are the amount you pay for equipment and other expenses. Similarly, rent is determined by the amount you need to pay each month in rental costs.

* A Good Place: A great location is one that is surrounded by a broad audience of people looking for services or products to buy. It is important to ensure that there is enough customers that your product or service is able to get to because it will allow for increased sales. Ample Parking: If you need to find the new area, you must be aware of whether or not you have enough parking space. Are the facilities big enough to hold all the equipment? In addition, if there's no parking available, you'll never make any sales! Limit your potential clients' visits to your location and don't let them go until they purchase something.

72

"Location Signage" To draw clients, you need to be able to clearly and easily identify your location within your vicinity. Your site must be easily accessible for motorists and pedestrians. Since, no one would ever want to travel one mile from their home to get a cupcake!

Proximity: Proximity measures how far you are from your site and the market you want to target. As closer you are, the easier customers will find it easier for your customers to purchase your product or service.

* Buyer Loyalty What is the loyalty of your customer base? How likely do you believe they'll continue to purchase from you? It's vital to understand this as it can help you decide the likelihood that a targeted market is ready to buy your product.

The Corporate Culture company's culture is developed by customers, employees, and even the general public. This includes dress code, manners of speech in the workplace, and rules. All of these elements make up a culture at work that creates a distinct brand for your company. Consider what type of place will help to make your company more prominent in the community. For instance, a building with lots of windows or one with plenty of parking spots for customers.

Be aware that location is the most important factor in the financial viability of vending machines located in public spaces. Take all the aspects above in mind when selecting locations so that your business can be extremely profitable.

The Best Locations

The vending machine business has grown exponentially because of the advancement in technology and the fact that people are working at a distance from their homes more often. With the ever-growing amount of locations that require vending equipment, it may be difficult to determine the best place to put them.

Vending machines sell a wide range of goods, such as drinks, snacks and cigarettes. This implies that they can be located nearly everywhere; However, certain locations provide more chances over others.

A crucial first step is to ensure you've explored all the options that are available. It is essential to find out whether your products are working well in the current areas, and if you're not sure which location they're purchased from there are options to determine. Machine locators for vending machines provide comprehensive details about each machine that is operating in your region, which includes precisely where it's located and the length of time it's been in operation. This tool can be used to identify machines that aren't making profits, which makes deciding the best place to put them significantly simpler!

It is also important to consider competitors when you decide the location for vending machines. It is important to choose places that give your business the highest chances of earning profits. For instance, if the majority smoking machines can be found found in middle schools, it might be worthwhile to put your brand new machines there too.

Students in high school are more likely to spend more on vending machines than other age groups and, consequently, will most likely spend money at the closest location to them.

An organized approach will also assist you in deciding the number of vending machines you will need at each location, so that you don't end up by having to many, or insufficient. It is recommended to install three vending machines for every 1,000 square feet in the largest locations. If you're thinking about a space that is more than thousands of square feet you'll be required to make sure you have enough machines in order to serve the amount of visitors who visit. Vending machines are available everywhere these days but some places provide better opportunities over other locations. When you take a strategic approach and weighing all your possibilities prior to making any new investments and making new investments, you will find it much easier to figure out which are the best locations to be for your company which will give your company an excellent chance of success!

How to Choose the Perfect Location to Vending

If you're working alongside others on a venture it's crucial to learn what you need to consider when choosing the right location to sell your products. This will help you come with the ideal location for your frequent customers because you'll know that they'll be satisfied and will never leave.

1. Determine the numbers First thing you need to do is determine the balance in your bank account

along with monthly expenses and revenue. If there's something wrong in these numbers, your company will be in a lot of trouble. Check to see if these numbers are accurate, as if they're correct, there's no reason you shouldn't be able choose the most suitable place to vend.

2. Find a location that is suitable when searching for a place ensure that the location is accessible. If you're considering placing your equipment in an office and you want to be able to access it, locate one that is accessible!

3. Learn about your market: You must examine your market and make sure that there is sufficient demand for your product or service. If there's not a demand, then you'll not be able to sell anything! After you've researched the market and determined that there's a sufficient demand, select the ideal place to vend. It must be located near enough to be easily accessible by sufficient customers who are likely to buy the products you offer.

4. Choose the most suitable place to sell your products in light of the above factors. If your business isn't generating profits it won't last for long.

Chapter 5: Selecting The Best Machine For Your

Business Of Vending

They are rather versatile equipment that can be used in a variety of environments from casinos to offices. If you're running or considering starting your own vending company selecting the appropriate machine for your needs is a crucial choice you must make.

This chapter outlines the most basic kinds of vending machines, and examines their application in both large-scale and small-scale household usage. The chapter will also discuss different factors that influence your decision-making to help you move your company to the next level.

The Factors that Influence Your Choice

After you've decided on the type of vending machine you wish to buy, you will be able to consider aspects that impact the cost and quality of the machine. There are a variety of factors to be taken into consideration when selecting the right machine. Certain aspects are specific to your specific situation and others are universally valid. It is important to consider these aspects when choosing the best vending machine for you:

Cost

Cost can be among the most crucial factors to consider when choosing the right vending machine, particularly when you're just beginning to get into the market. It is important to pick one that falls within your budget but still offers an experience that will please your clients. Higher prices do not necessarily mean better quality machine, and you have to strike a compromise that allows you to obtain the best worth from your investment.

Quality and Variety

The quality of the machine is equally vital, particularly if are in an area that has an influx of people. It is worth buying a commercial-grade vending device since they last longer, can handle greater weight and experience lower issues when compared with conventional equipment. The range of options available to the vending machine you choose to use is crucial because it lets you to satisfy the demands of every customer. You must be able to provide items that meet the requirements and desires of all who walk through your doors.

Size

The size of the machine is also crucial. Be sure to get the proper size to fit within the space set aside for it, while offering plenty of storage. Large machines are convenient however they take up a lot of space that's not being utilized. Even in smaller settings it's important to be aware of this as it can take up lots of space. Smaller machines tend to be less expensive however they are usually less efficient since they don't have sufficient parts or storage space to accommodate larger amounts of products.

Volume and Flexibility

It goes to the forefront, but quantity and adaptability are crucial aspects to take into consideration when choosing the right vending machine. It is essential to choose the machine that can handle large quantities of items and can fill multiple orders in a variety of places. This makes the machine more effective. A lot of machines let you change your order when it's full. This is very beneficial for you as it allows you to monitor the inventory. If changing products is not an option for your business, you need to look into refrigeration or temperature-controlled products that will keep the items cold without affecting their quality.

"Choosing the best Machine for your company" suggests that you should determine the kind of business you operate to be able to select the right vending machine that you can purchase. If you run a big office that is able to serve, it is logical for you to purchase the conveyer Vending Machine. If you have small-sized business that has less customers and fewer customers, it is possible that you can choose the Countertop Vending Machine will suit your requirements perfectly. The machines have to be purchased from supermarkets as well as warehouse stores. Therefore, they can only be purchased from third-party vendors like Costco. Consider the cash flow required to cover the cost of vending machines frequently prior to purchasing any equipment.

Vending Machine Sizes

If your business requires more than one machine, it's crucial to determine the size machine to purchase in accordance with the number of individuals who require service. It is possible that the Countertop Vending Machine may be an ideal choice in the event that your office only requires one machine. However, should you require multiple machines, you'll be better off using an e-conveyor Vending Machine.

Vending Machine Designs

For a smooth operation of a vending machine, it's crucial to look at the different models that are to choose from. The primary difference between two kinds of machines is the way they operate. They are both operated differently. Conveyor Vending Machine requires that customers pay and deposit money into the machine each time they're looking to buy something. In addition Countertop Vending Machines work differently. Countertop Vending Machine works differently by permitting customers to browse the internet and wait for items to be displayed before them.

The Machine's Place

You need to decide the place you'll put your vending machine before you decide on the model. The dimensions and style of your workplace will be essential factors to take into consideration. It is also crucial to figure out how much is required to run the machine. After you've decided on a machine then it's time to decide the location you'll be putting it. This will influence the number of patrons who visit your vending machine, specifically when they already purchase their food or drinks from somewhere else.

Vending Machine Things to Consider

There are a few things to consider when selecting vending machines, however, among the most crucial aspects is where you plan to place it. Customers should be able access the vending machine with ease which is why you should attempt to put it in an area that is crowded near the entrance , or somewhere that is easy to access.

Another aspect to take into consideration when selecting a vending machine is the number employees you employ. This will assist you in deciding the size of your machine. If you have fewer than five, the countertop model that is smaller in size will be sufficient. However, If the number of employees is greater than 5, it may be necessary to invest in an e-commerce-style conveyor treadmill. Also, you should consider the amount of cash flow that you have. If you don't have lots of money in a single time, the standard vending machine could be the best option for you.

Vending Machines In An Office Setting

They are typically placed in offices because they are convenient for customers and employees alike. They can be found both inside and outside. They are typically located at the entry point of a building because the majority of people have to pass by it throughout their daytime activities. When you think about your vending machine, the primary factor to consider is the location. It can impact the flow of customers and can cost your business if it's not in a convenient location.

Installation Questions

If you've made the decision to buy vending machines then you must consider where you'll place it. This is because clients often have to shell out lots of money buying food or drinks and can be risky if you're not cautious. It is also important to be aware of the space your machine is taking up. Be cautious when using machines with drinks or food items because it is possible to run the chance of boiling over or spilling hot drinks on employees and customers alike.

Machine Alternatives to Vending Machine
Alternatives

You might already have vending machines in your
workplace and are looking to upgrade it or find
one that is more suitable to your requirements
and requirements, which means it's the right time
to change to a different type of machine. It is the
Countertop Vending Machines are effective in
small offices, however they can be costly to
maintain when you have a lot of employees. The
best option is one called the Conveyor Vending
Machine that doesn't require any money from
customers to purchase products.

When you purchase new equipment for vending
machines there's always the risk of investing in
things that aren't necessary or might cause
problems in older machines. The best method to
determine the requirements is to make your list
of what you'd like the machine to accomplish for
you, and then begin getting rid of the ones that
don't perform. It is also important to take into
consideration your budget. This will allow you to
choose which kind of vending machine to best
meet your needs.

A lot of people aren't acquainted with vending machines and might be reluctant to handle them. If this happens at your workplace then you might be required to put up instructions in the back of every machine, telling customers what they can expect from them. These guidelines will ensure that your employees as well as customers are aware of how to operate the machines in a timely manner. It is also essential to note that many vending machines are built for use with coins. It might be easier to purchase machines that accept bills to decrease the amount of coins you lose at your workplace.

Chapter 6: What Are The Needs Of Customers In

A Vending Company?

The business of vending is extremely lucrative. However, how do you decide what you should sell to earn the most profit? It is important to think about the needs of your customers to ensure your vending machine business is able to grow. Here are a few things to consider:

Personalized service Make sure you offer a one-on-one experience in which customers get what they require and find out about the latest products.

* Location: Think about the place of your business. Are you in an area that is crowded? Consider the ideal location of your vending machines.

* Don't have too many machines. You're trying to draw customers and therefore having a small amount of machines in one location could be better.

• Try foods first Because you will need cash to keep your business going Begin by selling food items before working toward bigger items like computers or electronic devices. If you're planning on selling any kind of product ensure that your product is well-known and readily accessible (such such as computer).

* Take into consideration junk food: If are in search of pre-packaged, high-profit goods, you should consider snacks as a possible alternative. You should consider stocking up on when you are just beginning your journey, make sure you don't have too many machines stocked with the wrong items. Start with a small number of machines and make changes when you notice your level of success rising or decreasing. It might be worth buying more food items rather than other items to begin with because they're more sought-after and are easier to make profits from.

• Price your products accurately: Ensure all your prices are in line with the correct price. If your prices are set to low it could result in no profits or a losses, rather than bringing in revenue. If the price is set too high, buyers will have difficulty justifying buying your product.

* Make the most of sales on seasonal items: Prices for many items are typically cheaper during

the holidays. This is a great occasion to sell sweets as well as candy and snacks.

• Consider earning profit: You could consider making your products more valuable to make them more attractive and thus more worth it. For instance, if you could add an infuser bottle or a carbonated water dispenser, those upgrades could generate a lot more revenue than selling soda by itself.

* Discounts on bulk purchases If you can avail bulk discounts on essentials like cleaning products or food items This can significantly increase your profit margins as well as operating costs can be reduced.

You might want to consider investing in buying vending machines that you can sell in the event of selling your business.

Take a look at new technology You could consider the addition of the digital marketing and shopping cart devices to the vending machine to help you become more efficient and up-to-date with the latest trends.

* Put money into your equipment: Insuring money in better equipment like an electronic cash register is a good idea. This will improve the

customer-satisfaction level of the customers when using your vending machine, which ultimately makes for a lot of extra profit.

* Think about cash only If you're a retailer with a lot of customers who want to purchase items quickly you should consider making the vending machines only cash.

• Invest in advertising There are a variety of ways to promote your product to draw the attention of customers. You might want to consider purchasing digital marketing tools that can be utilized within a budget. Consider adding attractive signs that are accessible to anyone who walks by and will attract the attention of and possibly attract more customers.

People require water The addition of a large water dispenser will always be profitable for any vending machine company.

* Consider investing in a variety of products: Most people don't want to buy more than one product from your shop at the same time. Consider selling a variety of food items, both small and large products, or anything they could take home with them.

90

* Make investments in larger items: You can always think about expanding your business and installing more vending machines in various locations. This could increase the potential value of your company.

* Bartering: If you are short of vending machine parts it is possible to trade with businesses in order to procure the items you require.

Take a look at the convenience of pre-packaged food items. Food items that are pre-packaged, such as water bottles and fruit snacks are easy and simple for your clients to purchase at a vending machine.

You should think about using social media. Social media is a popular method to connect and exchange ideas. It is always worth considering making use of social media in your marketing strategies to grow your company's reach.

* Develop an application: Creating an application for your vending machine company is a great method to attract more customers.

Take into consideration the weather. Consider how the weather could impact your vending machine company, particularly that are too hot or wet out. It is possible to shut down your machine or at the very least reduce the amount of machines you've got on hand to ensure your investment is secured in the event that something goes wrong.

Don't purchase cheaply When you purchase items for your vending machine company, don't buy the most expensive item. Most of them are of poor quality and could leave those who purchase them sick. This can leave a negative impression on prospective customers, and it is better to stay clear of it completely.

Be mindful of hygiene As a seller, ensure that you keep all equipment clean, and in compliance with standards so that you don't encounter any issues with authorities.

Vending machines are able to go everywhere you can put vending machines virtually everywhere, making it simple for customers to purchase at your store anywhere.

Take into consideration the seasonality of your business Consider seasonality: Seasonal items are always worthwhile to take into consideration. It will help to keep profit flowing and increase the profitability of your business.

• Consider security investing in security cameras could help safeguard your business from potential security issues that could result from the cameras.

If you've got the funds, think about purchasing digital marketing tools that can be used to advertise your product at a lower cost. You can also engage individuals to sell your products on behalf of you in the event that you have more cash.

* Establish an amount to each machine. Because there are a variety of degrees of popularity for various kinds of vending machines, it is advisable to determine prices according to this information. If you're going to face lots of competition and similar machines, it is possible to set prices that are higher than if you only possess a couple of machines.

* Consider being selective: Consider being selective regarding vending machine products. This can make your business more profitable , by providing more customers.

Create the stock rotation schedule One way to increase profits is to create an inventory rotation schedule to ensure that you don't own lots of inventory that's not being sold efficiently while at the same time. This can help your company remain in business and minimize any losses in profits that could be incurred.

It is worth considering investing more If you are convinced that vending machines as well as other related products are profitable for you, think about getting a business credit or business loans. Make use of sales receipts Make a list of every sale you make and keep a log of these items in order to offer your customers receipts when they request. This can result in an increase in profits for your company and prospective customers who reach out to the manager or owner to buy products from the business.

Be aware of what your customers value When deciding what products to offer from vending machines, keep in mind what your customer is looking for in a vending machine. If they're particularly concerned about privacy think about not installing these kinds of machines in their homes or work place.

• Consider the location It is important to think about the area of your machines prior to making a decision on the machines you want to place in your company. If a lot of people walk around, you may be thinking about setting up more than one in specific locations.

Be aware of the risks to your safety There are risks with vending machines employed in the business, and health issues that can arise from it. Be sure to are equipped with the appropriate safety equipment and the proper procedure in place to ensure that you do not face any issues with authorities or with customers.

Take into consideration the surrounding area: You should think about what kind of location you'll be operating from when creating vending machines for your company. If the location is many foot-traffic it is possible to think about setting up more than one up in various locations.

Take into consideration taxes on sales. You might think about setting up federal and state sales tax on all your items. This will ensure that you are in compliance with your tax obligations and making sure that your customers are paying taxes.

Take into account customer service: If you've got a positive reputation for customer service this could be beneficial to invest more money on social media since this can help in getting more customers to return to your company.

Take into consideration marketing: Marketing is an essential component of vending machines. This will help you grow your business and ensure it's profitable over many years into the future.

Chapter 7: What Is The Cost Of Vending

Machines Cost?

There are plenty of excellent vending machines available these days. They provide the selection you desire and the cost you want including chips and candy. What is the cost?

In general, vending machine costs typically start at around 1,000 for the first machine. It could range from the most sophisticated drink dispensers, to basic snack machines. A recent report about CNN Money reported that one vending machine model cost $27,000. However with the advancement of technology and consumers demand more features for their machines, prices are always trending down however they are still expensive due to the consumer sensitivity to price.

They vary in their dimensions, which affects costs since larger ones are usually more costly. The dimensions of vending machines generally are four feet wide up to eight feet long and can range from three up to 5 feet high. These dimensions are merely an estimate and may be different based on the model you purchase.

What is the price vending machine prices cost? If you're searching for an efficient model within the lower price bracket be prepared to shell out between $1000 to $2000 to purchase a machine that offers snacks and holds up to 20 pounds of food. There are modern models that have electronic coin mechanisms and advanced snack dispensers for around $7000 each.

There are three main types of vending machines that are sold these days. They are soda and snack machines that are usually found in offices and break rooms. Coin-operated models tend to be more costly and can run as much as $20,000 per. They typically offer energy drinks, cigarettes or lottery tickets. The third kind or vending machine that is available for sale are combo snack/soda machine , which sells similar items at a higher cost. Combination models cost between $30,000 and $38,000 depending on the size and include features such as speaker, bill-dispensers as well as digital displays.

If you're considering investing into a machine for vending buying one made in the United States can cost more. vending machines made from China and Japan can cost upwards of $40,000 each. The cost for the Japanese vending machine is $30,000. You can also buy expensive-looking models that are made in the US made of high-end materials and solid workmanship at the range between $6,500 and $8,500.

It's not so much the size or model of machine that determines the pricing as much as the degree of automation. It's all about the ease of use for customers as well as efficiency to business executives. The more automated the machine more efficient and faster it is to use and more expensive the cost.

Generally, vending machines are an easy way to get beverages and snacks without having to travel far. The most effective models can be found in gas stations or convenience stores to facilitate impulse purchase at the convenience of your mobile or easing queues at cash machines on busy days. Vending machines can be a great option for businesses who want to increase their revenues while also providing a vital service to their employees and customers.

In a nutshell In short, vending machines have been the rage for a long time. However, with the advancements in technology, such as electronic coin mechanisms as well as digital reader for smart cards that allow music to be played and play music, this trend won't go off anytime in the near future. In general, the greater automated a vending machine is, the more expensive it will cost. However, these machines could be able to pay for themselves in a short time when properly maintained and operated. Vending machines are low-cost and profitable investment.

The cost of vending machines is around $1,000-$2,000, and that includes all the necessary equipment and labor to put it up it. It is vital to keep in mind that no vending machine created alike. They are available in a variety of designs and sizes. Certain models are more costly than others , based on their the quality, size and level of automation.

You might be searching for a vending machine for soda or snacks to make it easier to purchase impulse items in your break room at work it is possible to find one around $1,000-$2,000. If you require something more sophisticated , that will accept debit and credit cards, the cost of vending machines usually will increase dramatically. Certain models can be as high as $38,000! If you're searching for a machine that is stocked with exotic goods either at the convenience of your home or retail store, be prepared to shell out between $6,500-$8,500.

This table shows the typical cost you can expect to shell out for a simple vending machine that does not have many bells and whistles or even automation.

Model Price

Purchase a vending machine for sodas or snacks $1000-$2000

Purchase an automated vending machine for $3500 to $7000.

Get a cigarette, snack machine that can be rented for $8000 to $20,000

Get an exotic-looking vending machine between $6500 and $8500.

Chapter 8: Tips To Become Successful In The

Vending Business

A profitable vending business requires a variety of qualifications and abilities however, they are simple to master. If you're seeking to make money providing drinks or food for your clients, here's what you should be aware of.

The Vending Business

Since the beginning of the 19th century the market has grown into a flourishing business when street vendors started selling delicious food from carts along the city's streets. Since then, the market has expanded to higher-margin firms that sell through outlets like food and grocery stores. This chapter will examine the characteristics of successful vendors as well as provide strategies to be much more effective in the area.

Qualities of Successful Vendors

For the most out of your vending business, you'll need vendors who are comfortable with their work. If you're hoping to be successful, you must select the right vendor and assist them in

102

becoming better at what they do. Here are the numerous requirements that successful vendors be able to meet in order to make money from their vending businesses.

Skills

The ability to serve customers with refreshments or food is an essential requirement for a successful seller since it cuts down on time, helps people sleep better during a hectic day and also reduces the amount of turnover. It is essential to know how effective your employees are in opening for business often. Only way to achieve this is by training and experiences.

Communication

Since you are the one in the control over your workers, well-known vending businesses require the best communications skills to help your business grow. The way you train your employees shows your customers that you're professional and look to protect their interests. This is the reason why so cities with high-performing economies have successful vending enterprises: they've learned the art of communicating and know what customers require.

Attention to Attention to

The majority of people who run vending businesses utilize mobile apps, however they're not all doing it correctly. If you're a successful manager, you must demonstrate to that your staff members how to upload items into the app, make daily reports, and ensure the highest quality of service. It is important to demonstrate to your employees that paying attention to details is crucial to the success of your business.

Leadership

If you'd like your business to grow the best leaders will take their teams further than they could by themselves. It is essential to have an effective leader and employees who are willing to do whatever it takes to allow your vending company to be profitable. To be able to lead others you must be educated and this can make them more eager to work with you, and will allow the business to grow.

Negotiation

To be successful as a vending proprietor Sellers must be negotiating with their business partners, clients or customers. In certain situations you may need to collaborate with clients or other vendors to achieve the best results. The ability to negotiate well is essential for vending businesses to be successful.

Planning

A written plan is essential for any business that is starting The same is true to vending machines. It is essential to develop an outline to help to manage your business and start starting on the right path before the opening day. Use an example of a outline to help you plan for your business ahead of time.

Flexibility

There's a reason vending firms existed in the first place. They exist in order to offer customers with quick service. However, it is possible that opportunities arise which make the most profitable one appear less lucrative. If you're working towards making money the business you run must to be flexible enough in order to permit growth when the changing economic or business trends.

Entrepreneurial Spirit

Businesses that are successful don't just appear overnight. They require many years of hard work behind their backs and require constant maintenance in order to stay active and profitable. The owners of successful vending businesses need to have an entrepreneurial mindset to succeed and grow over time.

Education

If you are looking to be profitable in your vending industry you need to be aware of the basics of what you're doing. This training will differ for each business, but a wealth of options are available online to people who want to learn more about the field. If you are planning to launch your own vending business and are looking for the best strategies and details is essential. The information contained in this book won't make you successful in the field you choose but they will assist you begin.

Benefits of Being a Successful Vendor

The benefits that successful vendors offer their clients are many. The business of vending is both a physical or financial commitment, however it's well worth the possibility. Here are some advantages that successful vending vendors have over their customers.

Customer Service

A reliable vendor offers high-quality items in their stores with low costs and this is what customers appreciate! They love their favourite vendors since they find everything they desire in a short time. Customers love the fact that they can purchase from vending machines with less time than normal and are delighted because they don't have to stand in line during the day of their busy. The top vending establishments offer a guarantee of service or a type of refund if you're not happy with your purchase.

Convenience

Customers love the convenience of finding everything they require at one location whether it's snacks or candy for their business or family. The fact that sellers sell everything at one time will mean less time spent traveling across multiple locations and more money saved as you won't need to go out several times per week. It is essential for many peopleand buyers are always grateful to the sellers who offer them the convenience they require.

Affordability

Most vending machines are reasonably priced, particularly when compared to purchasing them from a typical retail store. The reason that a lot of businesses have vending machines within their establishments is due to the fact that it provides an easy access to food and beverages for employees and employees in the field. If you run an enterprise vending machines can be beneficial to save you money. You can purchase snacks from a vending machine and not worry about who you send out to or even providing food for your employees.

Value

The cost of vending machines is set by vendors. Vendors have the ability to set the price of products and change the prices to lower their prices to the general public. This benefits clients because it saves them money. This helps them save cash if they purchase products regularly and continue to use vending machines. Many people live off their vending machines. This is the reason why people like sellers who are transparent with their pricing.

Smaller business owners must recognize the advantages of running vending company. Understanding these advantages will make it clear the reasons why business owners enjoy having this venture at home. It is a fantastic option for business owners who want to serve their customers, market their items, and earn money.

Chapter 9: Does It Easy To Manage Vending

Machines?

Are you able to handle vending machines? If yes, what are the most challenging aspects of your job? And how did you conquer these challenges? Management is an advantage that is offered to certain professions. The more sought-after the position is, the more complicated of a job it's. For instance, the management of vending machines that sell food and drinks isn't easy to manage efficiently.

Business Management Strategies for Your New Business

Here are some helpful tips for managing your brand new venture in a manner that is beneficial for you as well as your business.

Tip 1: Be in contact with the customer

In relation to vending machines, it is important to keep in touch with your customer to know how the machine is running. If there are any issues you can go back and try things again. Make mistakes and learn from them, make notes and refine your strategies. For instance, if you do not keep your customers happy and satisfied, they might not come back for the next cold beverage or sweet treat. Make sure you report a high volume of transactions to show the cost of your plan and continue growing your business with ease.

Tip 2: Create A Solid Relationship with The Clientele

If you've built a strong relationship with your customers and they are more likely to come back and you'll have an easier time making money. Building this type of relationship may be difficult however it's worth the time and effort into it in the event that it pays off for you.

Tip 3: Determine What is the most talked about

You must ensure that you pick the appropriate vending machine for your location. Be sure to select a machine that is popular and will draw loyal customers for regular business. When you are selecting the items to put in the vending machine, it is important to ensure that customers will be interested in them and are willing to purchase the items at a reasonable price.

Tip 4: Select the right snack and beverage Choices

This is an essential element of your task. Choose a variety of your favorite drinks and snacks to ensure that your customers will always have something they like whenever they return to buy more. It is also essential to be aware of the calories in every item you place in the vending machine. For instance, if a vending machine is offering 300 calories in the snack options it may not suffice for today's healthy person who normally burns 1000 to 2000 calories in the course of the day.

Tip 5: Ensure You Keep Track Of Inventory and Sales Frequently

It is essential to track inventory and sales. Be sure to keep an eye at the ball to make sure that you're on top of your financials and operating a profitable business. Keep track of things that become popular or not to allow you to alter your decisions according to these results at a later time.

Tip 6: Ensure That the Vending Machine is always stocked

You must ensure that you have enough items available at the vending machine regardless of what time of the time it's. In this way, you can not be underselling or being unable to replenish stock, suffering a huge loss for both yourself as well as your business. If you are able to keep an adequate amount of stock available, you'll never be in danger of your machine becoming full or running out of stocks at a crucial date.

Tip 7: Create A Routinely and Systematic Inventory Management System

Management of inventory is a crucial aspect of the job. It is essential to have a method of monitoring inventory to ensure that you do not expose your business to losses that are not necessary in the event of to running out of inventory at the wrong time. Systems for inventory can be managed in various ways, and are generally easy to control. It is just a matter of making sure everything is in order and accessible in order to avoid any unpleasant circumstances.

Tipp 8: Always Keep Cleanliness and hygiene

It is an essential component of the vending machines in many. The merchandise within the case need to appear appealing, not like clean and unwashed displays customers simply give up on right away. Be sure that everything is well-organized and in order to ensure the best customer experience.

Tip 9: Create A well-informed sales team

You should ensure that you have some employees that are knowledgeable about the products or services, so that they can provide excellent customer service, rather than being unprofessional at the wrong time , and creating problems for customers due to how they handle their customers. A knowledgeable employee understands what customers want and when they're looking for it, and the amount they'd like it, which makes them an asset to your business.

Tip 10: Keep a record of your inventory Periodically

It is crucial to take this step to know which people are selling items and what everyone's position is with regards to what they're selling and the amount of it has been sold. If there is a problem in the machine it will be possible to identify the problem before it gets really bad and repair it rather than suffering for a long time due to an issue which could have been easily addressed should it have been discovered sooner.

Tips for Management 11: Be more generous In Your Financial Relationship With Your Employees This is something that you need to incorporate into your company's social norms. You should ensure that you provide benefits to your employees such as medical insurance and paid holidays as well as free meals. In this way, they won't receive incentives that they do not have and are more likely to purchase items at your vending machine if they feel that they are benefiting by having these benefits to them.

If You Have A Vending Machine Maintain The Customers Happy

Every busy person wants to have a fully-stocked snack machine in close proximity for quick access. Many people also appreciate the ability to get drinks and other snacks during breaks or for lunch during the daytime. Vending machines with snacks or drinks as well as an area to refill drinks they're drinking are extremely popular as an addition to the workplace vending machine. Here are some helpful tips to ensure that your vending machines are operating at their peak performance.

Tip 1: to know what people want the most often or when they Are Looking for It.

The most efficient method of doing this is to keep an eye on the purchases and sales at vending machines over time or by conducting surveys. The results of surveys will provide you with a clear idea of what the customers enjoy most and what they prefer about it most. This can help you determine which products are selling the most. For example, if customers prefer their drinks cold or cold, you'll be able to have ready-to-use coffee and tea.

Tip 2: Provide Something for Everyone

Be sure to offer a wide range of products to please the widest range of customers as you can. For instance, there are people who do not like coffee, and would rather drink tea instead. In the event that there's two types of coffee on vending machines, people might not be pleased because they might prefer tea and not purchase one because they're both exactly the same.

Tip 3: Buy An All-New Snack Machine Every couple of years

They contain lots of soda, with a short shelf time. If you plan to offer your customers a wide selection of fresh snacks and drinks You must ensure that you purchase an updated vending machine twice a year or less. This will increase your earnings since clients are always looking for new items and other items that they can make use of in their offices.

Tip 4: Provide Snacks For Meetings And Parties
This can increase profits for vending machine companies as snacks are offered when there is an absence of customers entering the building or office. It is also possible to see the sales increase in these instances if you are planning to attend meetings or have already attended an event where they have to consume something smaller such as chips, peanuts or popcorn. This can increase your sales, particularly if you're having unproductive days.

Tip 5: Buy Your Drinks and snacks from a Company That Offers Excellent Products
Since vending machines are designed to store and distribute food and beverages, the products that are stored in them must be high-quality products made of high-quality ingredients. Making investments in these products will result in a better experience for the customer and this

means they will be more likely buy them, particularly in times of low profits, when they need to be increased at every cost.

Tip 6: Ensure That Everything is In Order

This covers the machine, all the things within the machine, and the surrounding area. Maintenance is an essential part of keeping your customers happy. It is important to let them know that they can have food whenever they want.

7. Keep A Clean Space Near Your Vending Machines

It is important to ensure you're selling top quality products so that customers will keep coming to return for more every time they are able to test your products. If there is a mess around your business that is littered with potholes, cars or other debris, customers could slip while taking their snacks or drinks out of the machine, leading to an accident. Cleanliness can stop this.

Tip 8: Experiment With the Latest Snack Ideas as Often As You Do

Customers want to keep returning at your machines on a regular basis so that your profits keep growing. If you are able to offer something new and different and unique, they'll be coming back for more within a matter of minutes. For instance, you could give a unique popcorn flavor that is not available at their machine vending. It will stand out and cause people to want to purchase from your machine instead of other vending machines nearby.

Tip 9: Offer Weekend Specials

Customers can also be enticed to return during weekends by offering them an incentive to do the same. This can increase revenues since weekends are usually a slow time for offices, specifically during lunch hours and in the afternoon, when employees require a quick snack prior to starting work. You should ensure that you offer something fresh and interesting that will entice people to give your drinks and snacks to try out during the off-hours.

Tips 10: Make Extra Small Snacks and Drinks

Another way to make more money regularly. If, for instance, you're selling a lot of tea and coffee and tea, you can add couple of snacks so that customers can grab a snack anytime they'd like without having to purchase the larger items. It is preferential to offer small portions of your product rather than making people return to takeaway whenever they desire their typical drinks and snacks.

Chapter 10: What Are The Most Frequent Issues

Faced By Vending Machine Operators?

In the past decade, our dependence of vending machine has been increasing. Most people are aware that it's an enormous task to run an enterprise that sells vending machines however, there are some aspects that people do not think about. The book I'll be discussing typical issues faced by vending machine operators and ways they can be prevented.

1. The possibility of a lack of funds

For most vending machine operators the first stage is solely based on selling of snacks. When considering an investment, the price of operation must also be taken into consideration. While it is essential to make snacks available and turn an income, if investment decisions are not carried out at the right rate, the business may not be able to sustain its own costs.

2. Poor Customer Service

Vending machines are among the few vending machines that have integrated customer service.

While the majority of vending machine operators are extremely responsive, there are some who struggle to get their machines up and running. For instance owners of vending machines might not be able to service the machines in a timely manner or at all.

3. Stable Expansion Efforts

Vending machine owners typically offer beverages and snacks. There's a possibility to extend the reach of the business by offering additional products such as frozen and hot food, however not every is able to do this. Vendors must consider what other products can be sold to boost their income stream. For instance, if the vending machine serves drinks and snacks it is possible to sell hot meals from the same place.

4. Poor Maintenance

Another reason that can cause vending machine companies to fail is bad maintenance. Vending machines have to be properly maintained and stock. The owner of a vending machine who does not keep the machines adequately stocked is likely to have difficulty making money. When it is taking too long for stocking the machines or inventory are not up to date. This could negatively affect the sales and the amount of revenue consumers earn.

5. Poorly managed responses

The communication between the owner of the vending machine and employees must be extremely strong. This is particularly important in times of growth for businesses since it could lead to problems with employees who aren't prepared to take on such responsibilities. While some business owners might not be a problem but others may struggle to manage the demands. If an employee fails to satisfy this standard and the owner of the machine is likely to face issues.

6. Inexperienced and inexperienced with Inventory Management

The tracking of inventory is crucial to any machine business. An organization that does not have proper inventory management could face issues with customer service as well as other employees responsible for the delivery and stocking of products. Based on the quality of that vending machine is stock and how well the customer service was managed, it could be affected when inventory of certain products is excessive or even low in a certain area or when specific times during the daytime are flooded with customers who want to purchase products.

7. Poorly managed Vending Machines

The most crucial aspect of running vending machine is the management of it. If the vending machine isn't operating correctly, it will have more risk of loss. This could include items that aren't being filled or maintained within the appropriate period of time.

8. Poor Location for Business/Vending Machines/Location of the Business

There are numerous factors to take into consideration when selecting the best location for vending machines. If the location is not well-known, has low visibility or is situated in an region where sales are likely be low, you'll struggle to gain growth and revenue for the company.

9. Unhappy Employee Relations

One of the best aspect of expanding your business is to hire employees that can manage duties and tasks. This is particularly true when an organization plans to expand to a larger location and plans to employ additional employees to satisfy the demands and wants of its customers. Many vending machine owners have a hard time finding the perfect match, so they are forced to hire people who do not meet their expectations or struggle to complete their jobs properly.

10. Lack of Innovation/Creativity

If you decide it's time to grow your vending machine business, you'll need to think creatively for it to be successful. Otherwise you'll be like the other businesses that have been unsuccessful over the years. When you are expanding to a new location it is important to be creative in your approach.

11. Poor Payroll Management

Another aspect of business that should be taken into consideration is the management of payroll. If payroll management is not properly managed employees could not have access to a lot of funds and may not be able to manage their family or personal costs. The cost of college tuition, car notes, repairs to homes, and many other things can burden those with no or little funds to pay for these costs. When businesses fail due to inadequate payroll management and management, they usually have to lose employees as well as customers.

12. Inability to see Clear Vision

A business that isn't equipped with a clear plan of action may fail to achieve its goals. If you don't have a clear timeframe within which you'd like to accomplish certain goals, chances are you'll be unable to meet your goals. No matter if you're working in a group or working on your own having a clear and concise vision is essential.

13. Poor Maintenance of Vending Machines

Unprofessional maintenance practices for vending machines could cause a variety of issues over time with regard to the overall performance and quality of the machines owned by the company. If a vending device is not properly maintained the money could be lost due failure of products or broken machines. When customers visit vending machine companies they want to select an item that is well-maintained and meets their requirements. A lack of maintenance can result in something as minor as broken glassthat makes customers feel uneasy about purchasing.

14. The Company Lacks Transparency

If you operate an organization there is a expectation of openness; otherwise there could be issues with your clients or the general public. Transparency in your business dealings is essential for any business which wants to be successful in the in the long run. If a business isn't transparent, there's an increased chance that customers will encounter problems with the company or their purchases.

15. Poor Advertising Practices

In the realm of advertising, numerous businesses are successful However, many businesses aren't successful. If you do not have a clear idea of what you would like to accomplish and the approach you'll take to specific marketing initiatives it may be difficult to properly advertise and get traction within your industry. If your business doesn't recognize the right ways to effectively advertise, there may be problems with overall growth and sales for the company overall.

16. Insufficient Customer Service

A variety of aspects need to be addressed in order to meet the needs of customers with respect to customer service. If you offer a superior quality of customer service and provide a high level of customer service, customers are more likely to come back to your services and products in the event that they are looking for something different and new. No matter what your product or service may be the customers would like to feel like they can trust your business, especially in the event that you can provide a high standard of service or offer them the highest quality of service.

17. Small-sized Company Size

If you manage an unproportional vending machine business it is likely of sales being lower. This is especially true when the people on your team are experienced and abilities to manage the business efficiently. If you don't have sufficient resources to manage your costs and sales this could strain your business and cause failure if you do not learn from the experience.

18. Failure to Set Goals/Risk-Taking

They want to be aware of what to be expecting from the company they work with. If a business fails to demonstrate the world that they have a mission and a strategy, that it is innovative in its method of operation, and will take risks to achieve its goals and goals, it may face major issues in the near future. Lack of planning and thinking ahead are typical issues for firms.

19. The lack of transparency

You must be transparent in your practices in sales and delivery processes to ensure that your customers are satisfied and feel like they can trust your company. If customers don't believe in you or believe that they can't trust your company's offerings or products, then there will be no long-term success. If a business isn't transparent, it is difficult for the public to believe in the company.

20. The absence of Quality Products/Services

A company that doesn't offer high-quality products or services in its offerings will have a difficult time gaining traction and keep its customers satisfied for the long haul. A vending machine business that fails to maintain high standards for its goods and services might find that customers are more likely to switch to alternatives similar to theirs with an established reputation. Customers looking for new products or services could look for different vending machine vendors if they realize that yours is falling short of standards in terms of performance or customer service.

21. The absence of branding and marketing

In terms of marketing, a business that doesn't properly market its own products or services could have trouble gaining recognition. When consumers are looking for vending machine items consumers need to know what's available on the market and what price, quality and other features each firm provides. A vending machine manufacturer with no brand strategy in place may have difficulty at retaining customers for a long time.

22. Poor Location Selection

Vending machine businesses typically put their machines in areas that will attract the most customers and achieve their objectives in terms of growth and sales. If you put your machines in unsuitable places, it could hinder the overall success of the vending machine business. If you place your machines in areas that don't look appealing and you're not doing it simply because you don't have a other option, customers are likely to steer clear of using the machines for their own needs. this issue could make it difficult for you to achieve your goals in business.

23. No brand Identity

If people find out the vending machines firm has no clear brand identity or identity strategy and strategy, they may be skeptical about its ability to provide quality products or services. If you don't have the name of a brand, consumers might conclude that your products or services aren't of high quality. If they think that your business doesn't have a image, they may look at different vending machine businesses which can provide the goods and services that you require to gain an edge over your competitors.

24. A Brand Without Advertising Strategy

You must have a strategy to connect with your target audience and sell your service or product effectively in terms of advertising. If you don't have a strategy for branding and an advertising plan, there's a high chance that your sales will be affected. If customers don't see your business offering top-quality items or solutions, they might turn to competitors for their needs.

25. Negative Publicity

If you're thinking of the idea of starting your own business but are unsure of the steps to be expected, it's useful to find reliable companies who can provide advice regarding the beginning and growth of vending machine companies This can aid you in avoiding the potential risks. If you're facing negative media coverage or through social media there are numerous ways for your business to be affected when you're not ready for it. If you're not ready to deal with negative publicity, and it damages the products or services you offer by any means, consumers will cease to purchase products or services, and you might not be able that you can rely on in the near future.

26. The absence of Consumer Education

A vending machine business must have a strategy to help customers understand the services and products offered by the business. If customers aren't aware of the nature of your services or products are, they'll conclude that they're not trustworthy or of top quality in addressing their needs. A vending machine business that provides its customers with inadequate information about its products may create real issues for its long-term performance.

27. Poor Employee Morale

If employees are unhappy and unhappy, they are likely to spread the negativity to customers, making it difficult to meet their objectives. If an employee of vending machine companies is not displaying a positive attitude and isn't fully committed to the objectives of the company, the employee could create difficulties for the company. Many believe that having unprofessional employees in the company that sells vending machines can be more important than an excellent product.

28. No Social Media Strategy

If you're not connecting with individuals through social media and presenting your company and its offerings, you'll be less likely to achieve your goals for growth and sales. Social media is a crucial component of the customer experience. Therefore, any company that fails to effectively engage with customers using these platforms will be unable to grow.

29. Poor Website Design

There must be a clear emphasis on the web's design in the case of corporate websites. Your website should be simple to browsethrough, and visitors will be able to locate what they're looking for on your website It is also beneficial to have some customer reviews on your site. If your customers are unable to discover your products or services on your website it's difficult to help them reach their objectives.

30. No Social Media Strategy

Social platforms on the internet are a great opportunity for customers to share their opinions about their experiences using your products and services , ensuring that you receive useful feedback that can assist you in improving the standard of your products or service offerings and help increase sales over the long run. If your customers don't have a strategy for social media and plan, they'll just go to the site of a competitor in search of what they are looking for of a vending machines manufacturer.

Chapter 11: How To Provide Items At A Low Cost

You've likely been in front of a vending machine. In the grocery store or in the cafe. Have you ever witnessed one deliver goods? Most likely not, since it's not there. It's not yet.

But that doesn't mean that you cannot build your own and then market it in the same way! This is what we'll examine in this section: how to provide products at the lowest cost by with a vending machine-based business model.

What is A Vending Machine Business?

To make this chapter I'll define a"vending machine" company as an (furniture) company that offers goods through vending machines. This does not include delivery services or drones or other methods, such as drones. The aim is to ship goods as quickly and inexpensively as you can. Customers are also not permitted to get in contact with the machines due to health and safety reasons.

On another on the other hand, one could claim that companies such as Amazon provide their goods through vending machines. Also, Post

Office. Post Office in some cases. Both businesses have been classified as vending machine business We'll be focused on the vending machine portion of their business.

There's no need to design your own machines and then market them in this manner. There is a possibility to sell products at low expenses through vending machines, when you already have the equipment you need and knowledge. Let's get into it!

The components of the model we'll review can be described in one word: cost-effective. It's comprised of transportation devices, or vending machines in themselves that are easy to put together, disassemble and operate. Some may consider it to be an unnecessary expense, but others see this as investment which can be worth it in the end. It's all about the amount of money you'd like to put into your business.

Product placement is an additional important element of this model of business. It is up to you to decide which location you would like your items to be delivered and what items will be most beneficial to potential clients or customers. You may also choose to focus on delivering a specific type of product, like, for instance, food items. You may also choose an appropriate price range or the quality of your products. This lets you concentrate on one item while other businesses take care of other items (and competitors). Although it's not mandatory delivery services that are available through vending machines can provide an distinctive, unique and (usually) less expensive service, which is the reason costs are generally reduced to an affordable level.

The principal goal of this model of business is to provide goods quickly and at a cost as low as you can while earning a profit. That's not enough to ensure that your investment is profitable however. You'll also need to determine what you'd like to get returned, for example, offering clients the option of paying for their purchase in advance or sending the order to credit card and asking for the payment following. If you opt for cash payments prior to the order it is likely that you'll get fewer customers as it requires that they'll have to put in more effort (and the question of whether there's enough money available).

There's no need for websites for this type of business. It is also possible to use Facebook, Instagram, Twitter and Messenger to send direct messages (DMs). It is also possible to use video calls to make the experience to feel more intimate. There are numerous apps which let you connect to clients via your mobile phone.

If we look back at our definition of business using vending machines it is possible to say it is a business that provides goods through vending machines. Due to its low cost and speed (delivery times vary based on the item) It is a good fit for businesses online across a variety of fields. It could also be applied to other sectors (such as distribution of books).

Take for instance the volume of books released each year. There is no reason to print and then distributing them via an established publisher. It is more efficient to utilize vending machines to provide them to you, or hand the work to companies which already offer this service through vending machines.

We've covered a variety of aspects of this model in depth However, there's one final aspect I'd like to discuss what could cause it to not be profitable, or what might be the cause of it. We don't have any idea of how much equipment that is cost-effective is available, nor what it does. We know only what it is and what it's supposed to be doing from its name. In addition, we don't know how much to put into machinery and the types of vending machines that are best suited to your needs. These are only examples but it could possibly be the opposite around.

What are the possibilities?

Vending machines are extensively used in a range types of businesses. The main aspects that determine the likelihood that this model will be successful in the long run are: first, how much you're willing to put into it, then the amount of items you'll be able deliver and, thirdly the variety of products you're able to offer. It is important to note that these aspects are also dependent on your abilities.

If it's your intention to get all-in by delivering diverse products, including footwear, clothing food, drinks, and even food in stationary (business). In this case you'll need to make an enormous investment in the beginning.

Chapter 12: Vending Machines Tips

Start your business by following these guidelines!

Tips #1: Choose the most appropriate vending machine to serve your needs. A majority of businesses choose the vending machine that is near their workplace or home. If you're unsure of what type you should choose you can try a couple of various types at different locations and find out what is most effective for your particular market.

Tip #2: Get a license! The local government will most likely require an operating license before you can sell products anywhere. The process of obtaining a license could take a while however the result is worthwhile in the end since it lets you become more organised and in compliance to health regulations, such as hygiene and food safety.

Tip #3: Employ employees. If you are planning to hire employees, you'll require a plan to meet their wages and have an area large enough to allow employees to work.

Tips #4: Make your own business plan. The business plan you create should include information about everything previously mentioned, including where your vending machine is and the arrangement you have with the owner of the location.

Tip #5: Get insurance! Insurance is extremely beneficial in the running of vending machine businesses since it allows you to ensure the security of your merchandise, as well as protect against dispute that may be triggered between you and a different participant in the vending machine business.

TIP #6: Register the company. Be aware that if you decide to incorporate as the status of an S Corporation, you will have to also pay taxes for employees. Check on the internet for specific tax issues.

Tip #7: Make an online presence. For entrepreneurs, you are essential to have a web presence. Creating websites for vending machine businesses is a smart way to inform your customers about the services or products you offer as well as information on where you're located at any moment.

Tip #8: Begin slow! Set yourself small targets and then move up from there. Don't endeavor selling thousands of dollars of merchandise in the first day. Start small and build up slowly.

Tips #9: Make sure you keep stock of fresh products available! If you are planning to vend food items, you'll want to make sure that your inventory is always current. If you don't replenish frequently, your the customers will stop buying your items, and then they'll be sold at a cheaper price by rivals in the market, or even taken.

TIP #10: Be sociable and positive! Your customers will be impressed by your confidence because many tasks must be completed behind the scenes in managing an enterprise. It is impossible to predict what will happen, so be confident and strong.

Tips #11: Buy items that are branded! If you purchase branded products instead of generic items it is possible to sell more products since people are less likely to buy something they've previously seen. This can increase your brand's appearance and make it more trustworthy to prospective customers.

Tip #12: Do your research on your market! Research is a crucial aspect of every business, particularly in the case of vending machines. It can be exhausting when you're reviewing sales figures every day Try to figure out an approach that works for you.

Tips #13: Use the internet to find potential customers! Most people don't realize this that the vast majority of vending company proprietors make a substantial percentage of their sales online.

Tip #14: Improve your sales! Explore different locations daily to determine which one is best for your service or product. It's impossible to get to know until you've tried it, so don't be afraid to explore a few places until you discover the perfect one.

Tip #15: Record everything. It is essential record all your transactions in order to file them with accuracy when you file your taxes at conclusion in the calendar year. It can be a hassle and time-consuming, however when you start from the beginning you can be able to keep all of your transactions in order.

Tip #16: Be committed to your business and stay with it! Don't forget that you are in business, and you need to be prepared for any hurdles that could occur. Be aware of the reason you started this particular business initially regardless of whether it's because of the excitement, for financial reasons, or because you like vending. It's a commitment of time and money as well as effort and you must remember the goal and be aware of the meaning before moving onto the next step.

Tip #17: Request assistance when you need it! Do not be shy about asking for assistance when you need it particularly if your company is relatively fresh. A lot of vending machine owners seek assistance from business owners. You may also seek out someone to guide you and teach you the right way to setting up and running your personal business.

Tip #18: Be prepared to tackle any issues that occur! Prepare yourself and be ready to deal with any issues that arise, including concerns about health regulation violations or theft. The issues could arise when there are many big companies operating within the same industry, or your products are well-known enough to draw competitors. It is likely that these situations will occur but if you're prepared, you will be able to tackle every issue that comes up and eventually be the winner.

Tip #19: Begin to purchase the machines! After you have located the location, decided on the quantity of product you want to sell and decided on your price then you can begin placing an order for the required vending equipment.

Tip #20: Be prepared to be a target for theft! The majority of vending machines are targeted by thieves who wish to sell the merchandise at different locations or grab them to themselves. There are now few businesses where the proprietor does not at least keep an eye on the business at least once a every day. If you don't monitor your company at least once a day, thieves may try to steal your business Be prepared for this , and ensure that you notify any thefts to the authorities in your area.

Tip #21: Manage your business in the right manner! This is so crucial that it is hard to emphasize the importance of it.

Tip 22 Be honest! If you're dishonest with your clients, they are unlikely to return to them. Even if it's one time thing that could erode your reputation as a trustworthy business , and more importantly your credibility in opinions of your clients.

Tip 23 Be aware of the competitors. Always ensure that you are aware of what's happening in the area and what other vending machines are offering to their customers.

Tip #24: Everything should be done in moderation! This is the time when we all have to step back and take a break for a moment. There's nothing more frustrating than burning yourself trying to market an item or service you might not be skilled at. This could cause many issues, stress, and even burning out. Burnt-out people tend not to work all the time and are only a little motivation.

Tip 25: Keep current with the latest technological advancements! Technology has made it much easier to maintain your business. It is possible to put your product on a site so that buyers can purchase them directly through your business, and you'll always be aware of how much profit is made. You can also buy social media accounts like YouTube, Facebook, and Twitter which will allow you to be able to upload images of your products using.

Tip #26: Start your own blog! It's not just great to keep your customers updated and entertained, it's also great to increase the search engine ranking. With your own blog, you can add keywords that will help people to find your website on search engines.

Tips #27: Be current! You must stay up to date with all the latest innovations regarding vending machine technology. Stay up to date with innovative products and methods people utilize vending machines. If you can do this, you will definitely keep ahead of the rest in your field.

Tip #28: Create an account in a different bank! Make sure your business money is separate from the funds you utilize for your personal. Maintaining this separation will allow you to invest in your business and obtain loans should you decide to increase your product line or acquire new equipment.

Tip #29 Tip #29: Be open to all kinds of products! This will provide you with more choices regarding the kinds of items you can offer and the amount of money you could earn from it. If a completely new market becomes available and you want to take advantage of it, then do so.

Tip #30: Stay up with the latest developments. If something happens within the field Don't be afraid to alter your strategy. Find ways to make the most of the new developments and ensure that you are up-to-date on the latest products and trends. It is an opportunity to increase your product range or If it's a better method to market your product and services, then do it!

Tip #31: Be aware of when it's time to stop! Some people believe they are able to sell every item they come across however, it's clear that this is not true. There's an appropriate time and place for everything, so be aware of when you have enough to stop things from getting out of control.

Tip #32: Pay attention listen to what your customers say! Focus on what your clients want and offer them precisely what they need. Don't be a fool and believe that you'll succeed since you know what you have do.

Tip #33: Be focussed and in the present! A lot of people are thinking about the future and a variety of things however the reality is that the majority of the plans they make will never be implemented while they're in the process of thinking about these things.

Tip #34: Set up your own budget! You should know exactly where you'll spend your money so that you don't have any unexpected costs or unplanned expenses.

Tip #35: Don't get too cocky! This happens often. Salespeople who promote your product or service attempt to be too confident but are then dismissed later. It's tough to put others down when you're still atop the mountain. If you ever feel like that take a moment to remember that a lot of people are working the same way as you do.

Tip #36: Never be afraid to make mistakes! You will make errors in business, so don't let it cause you to be scared. It is a good idea to use it to learn from your mistakes and try to succeed next time around.

Tip #37: Be engaged! It's easy to quit something that's not working, and look for something that is more lucrative. But, you must be motivated because it takes time before things start to work to your advantage.

Chapter 13: What Are The Future Opportunities

Of This Business?

We've found that vending has an opportunity in the age of advanced technology, thanks to the growing demand. People are becoming more conscious of what they consume, but time is a resource that people have a shortage of. Vendors are beginning to set up in schools across the nation in order to help kids who are either too young or unable to take time off for lunch.

This chapter will cover the possibilities of vending machines being used in your school, at home or other locations you think of where snacks could be required!

The future of vending machines are quite promising indeed. As people become more mindful of their health, however time is something people have a shortage of. Vendors have started to open their doors to schools across the country in order to help youngsters who are too young or are unable to take time off to eat lunch.

The possibilities of vending industries are endless and I can see it continuing for many years with the following methods:

Vending machines at home Adults and children alike will continue to seek snacks of this kind throughout their time off. You might even be able to purchase a variety of pre-made meals at these vending machines (i.e. for instance, if you have children who don't eat eggs).

• Vending Machines in public Places It is the government's intention pushing to make healthier choices easier to access for those who are always on the move. We've witnessed this in the form of Baby on Board signs in vehicles and now, at gas stations.

Schools are also aware vending machines are an effective way to make students more excited about eating fresh fruits and vegetables instead of eating junk food.

A report from the 3rd of February, 2012 states "One of the major reasons to bring Juicery in school is the fact that it eliminates the guesswork of encouraging kids to eat better. With the increasing number of children who are turning to junk food following the recession, and with a renewed importance on obesity in childhood this point couldn't be more crucial."

Someday, I wish this same method that is used in bringing food items to people will also be used for medical purposes. Imagine a vending machines in every school with vaccines on hand and you don't need to be checking for shots again!

"The Vending Industry The future of vending is extremely bright. Technology has made life easier and the companies that are that are aware of this will continue to prosper.

A few years ago the vending machine industry was in its prime. With people struggling with cash to make ends meet it was hard to believe that there was anything to hinder the growth of these stores that cater to everything you need, including food and office supplies.

However, over time, the rapid use of tablets and smartphones with their robust apps have threatened to destroy the long-standing business model. Could these new devices be the end of the road of vending machines? It could take a while before we can say for certain. It's important to pay attention to the trends within the field today to allow your business to adjust accordingly, if needed.

Vending Machine Industry Numbers

According to the research firm IBISWorld the vending industry earned $22.7 billion in revenues in 2012 all by itself. The majority of the revenue came from beverages and food items while other items made up the rest. Of the items sold, snacks and candy made up 35% of the sales while beverages made up 30% and tobacco accounted for less than one percent of sales.

The revenue for the year was forecast to have increased by 1.9 percent, which implies that the figure should be about $23 billion by the time 2013 is close.

"Continued expansion in the food and beverage sector will increase the demand for vending machines," Market research company stated. "Vending machines are vital for food companies since they let customers purchase essential products without having to leave their seats."

Vending Machine Development Around the World

The United States is still one of the top countries for vending machines. But, other countries around the world are beginning to see an increase in use of vending machines according to IBISWorld. The sales of vending machines were greater than $1 billion Australia and Japan in the year before and are projected to increase by more than three percent per year up to 2017. For Canada as well as Western Europe, annual growth is expected to be much higher at 4percent. Although this is a good thing for vending machine companies however, it's also causing some to come up with a brand new marketing tool to meet the needs of the changing demands of customers with their tablets and smartphones. "The food service industry is heavily dependent upon the consumption of food during the moment of consumption. Using vending machines for purchase can help reduce costs," IBISWorld noted. "This is possible by cutting down on the number of vending machines at one location, or by providing products from more than the same vendor."

Vending Machine Development Across Industries as well as Price Points

Although sales figures are the primary reason for this, some companies have suffered from the rise in consumption of beverages and snacks.

"Traditional store chains have historically depended upon hot meals and drink sales to generate sales," the market research firm said. "As more customers buy drinks and snacks from vending machines, it will decrease the revenue of these businesses."

According to IBISWorld The typical snack machine is priced between $2,000 and $1,500 on the internet or in retail stores. However, the vending machine business is seeing an increase in the number of affordable options for those looking to buy between $100 and $500. They are very well-liked in places where businesses are looking to provide their employees with the opportunity to buy cheaper alternatives to the offerings from vending machines which cost three times the amount.

It is also becoming popular for businesses to offer drinks and snacks through vending machines which are smaller than conventional vending machines. This makes it easier to access to the machines, which may affect sales, but is also intended to limit the weight gain caused by excessive snacking.

But, some of these latest vending machines offer higher earning opportunities than their less expensive counterparts, because they can offer more products within the same time.

Furthermore, because these machines are generally situated in cafeterias or restaurants they can be used to reward employees who adhere to healthy eating habits.

The Long-Term Machine Future Outlook

As new vending machines are available to be able to adapt to changing consumer desires, it's impossible to predict which future is in store for the sector. And as technology continues advance, we'll probably witness more machines being developed in the near future. But these advancements will improve and enhance older machines, or take them out completely.

We are aware that these businesses must continue to innovate to ensure that their products to stay current and popular. That means they need to research consumer trends , and then using this information to create lucrative opportunities.

"The vending machines that are small-ticket will be highly competitive in the coming years and manufacturers will be expected to compete in terms of prices, features, as well as the location of their machines," IBISWorld concluded. "Several businesses have joined the marketplace in the last decade, however just a few big companies account for a significant portion of sales."

Vending Machine Business Advice and Partnership Opportunities

If you're thinking of expanding your business to vending machine manufacturing. If that's the case the most effective methods to achieve this is to partner with other businesses who are already in the business. The trick is to discover the names of those companies in order to establish a relationships with them. Then, discover the ways they've managed to increase their revenue. Through these partnerships, you'll have the opportunity to take a share of the profits you typically earn from vending machines, and also offer these machines at different locations. This allows your products to be available to a wider audience who might not otherwise have access them, which could increase the sales and profits.

Chapter 14: Machine Types Machine Types

There are a variety of vending machines that are available to purchase. From the classic vending machine that has a familiar coin-operated mechanism, to one that accepts payments through credit cards There is a suitable vending machine for each business!

The first is the powered by manual, which are also called stackers or automated coin-op. These machines do not require coins, but they run with electricity. The staff utilizes levers and controls to place snacks in their hopper and distribute snacks when someone comes near.

Also, semi-automatic versions with a pre-programmed space for food items inside the machine. This means staff are not required to load these machines. Automated machines are often used to sell drinks and snacks and are able to only offer one type of item or a variety according to what the client would like to purchase.

They can be found in various settings like schools, offices supermarkets, airports or in doctor's surgeries. The price of these automated vending machines is determined by the amount of items they have and if they accept coins or not. Examples include the Coin-operated Vending Machine, which goes to the top of the list, given that you already know what they are.

The second type is self-service machines. It allows you to enter the machine and decide what kind of food or beverage you'd like. The machine offers you the option of choosing from a variety of choices, and if you select one that you like, it will stop making whatever it was processing and will give your selection immediately. Staff are not required for these machines as they feature slot machines for coins located on the right side. Examples include self-service vending machines, yellow Ticket machines in certain countries, and the Mini-mart Vending Machine.

The stock dispenser can be described as a machine that has only a few drinks or snacks in the hopper. These are automatically dispensable via an automated mechanism each time people approach it. Simply thinking about these machines brings us to thoughts of hamburger and hot dogs However, the advent that the web has brought has made it possible for the use of stock dispensers to serve a variety of purposes. Examples are Stock Vending machines.

The most popular kind of vending machine that we all have in common are the snacks machines. The machine we're familiar with can be found in schools, streets shops, airports, and even in shops all over the world. Nearly every city and town has at the very least one. These devices exist for a considerable period of time , and have been used in various products over their existence. They are frequently used in close proximity to toilets to decrease the amount of people who make use of them.

Another kind of device is the minibar. They are standalone gadgets that are available in bars and hotels around the world. The cost varies between countries It is advisable to research online prior to deciding to purchase one.

A very loved vending machines is the snack bar. It's an incredibly small structure that houses varieties of drinks or snacks. They are typically found in fast food establishments as well as supermarkets and other shops.

Self-service vending machines are often referred to by its alternative name, Automatic meters dispenser (AMID). This small machine features doors that open whenever someone walks by it. Inside, various items are generally stored in boxes or pads at specific heights. If someone is looking to purchase something at this machine, they need to remove the appropriate pad to get their beverage or snack, and then lift up the pad close to it, to stop the process, and then receive the purchase. These machines are able to be utilized in offices, schools and even in shops.

The vending machine can also be described as a machine controlled by coins that has its own tale to tell since it's easy to determine the purpose of these machines. They are among the most popular varieties of vending machines around the world They usually have drinks or snacks and require the user to place coins in an opening on the side of the. If they don't have money, they can not be able to purchase anything from the. A different type is referred to as a product dispenser since it could be filled with any number of drinks and snacks that consumers are able to select from when approaching it. The options could comprise an assortment of products like chocolate bars, drinks and so on and the machine can release the items by pushing them upwards. The vending machine that has change is a different type that we've seen in this guide. This is the most well-known variation of the same theme and they are simple to use since they require the user to put coins or money in. They're also known as change machines or coin changers and the procedure is straightforward as it allows the product to be released when you push an icon next to the slot that holds coins.

A different name to refer to a vending machines is a dispenser for merchandise. It is a device that supports used in hotels, pubs and supermarkets, as well as pharmacies. It is easy to distribute items as they are placed on shelves or placed in pads that are connected to the machine using the use of a sensor. If a customer comes near it the machine will pick the item up from the place where it's placed and take it away from the machine in the direction it is intended to go.

The vending machine that comes with vending software is a kind of vending machine which has been in use for a long time and is called a drum vend in some countries , and has been affectionately named coin dispenser. The machine is filled with drinks or snacks that are programmed so as to let the items to exchange money or coins. They are simple to use since they require the user to insert money or coins, and then turn off the device.

Another kind in vending machine dispensers is the one that sells products, located in workplaces and in the neighborhood. It could contain items that are of various kinds, such as tea, soft drinks chocolate bars, crisps, and snacks like sandwiches. Like other vending machines, this one is very simple to operate as it requires only you to put coins or cash into the coin slot, and then just wait for the products you wish to release from the location they're placed inside the chest.

The vending machine that is controlled by a computer can be described as a product dispenser that is equipped with computers. There are many of these machines around the world however they were originally constructed by Japan in Japan and United Kingdom mainly. They are very simple machines as they require only that you insert coins or money into their pad before you can use them. the account is created, and then the machine will let you have your drink or snack as you approach it.

The vending machine which serves frozen ice cream is a very popular kind of vending machine particularly within North America and Europe, where the ice cream is a significant element of the diet of many people. There are numerous kinds, however they are designed to be easy and simple to use.

The vending machine for ice cream is one of the most popular types because it is utilized in places where people love the taste of ice cream. Ice cream is available in machines that resemble those used to serve drinks and sweets, but it isn't always the situation. A good example is the ice cream dispenser. they are similar to drink dispensers, however, they serve ice cream instead.

Another kind of vending machine used to sell soft drinks as well as other items that people love to eat is referred to as a drink-dispensing device. These are typically located in hospitals, workplaces airports, public schools, and not forgetting bars and pubs. They also offer a variety of beverages that they are able to offer, including hot drinks and soft drinks such as coffee.

A beverage machine is a popular vending machine as it's widely used in places where people drink a lot. It's also easy since the client only has to put coins or cash into the machine to act like a bank account. As they approach, the machine will let them drink or other items stored inside the chest. You can also purchase alcohol-based drinks at vending machines. This is typically the situation when a bar does offer a drink but places one in its place. It is an effective method to get customers to purchase drinks since they're used to purchasing them through machines. An appliance that serves drinks is known as the drink dispenser. These are typically found in airports, hospitals, workplaces and even bars and pubs.

There are two kinds of drinks dispensers around for a long time; they are also known as coin-operated drink machines. They were invented within the USA in the 1940s, and have been in use since then, but their design has evolved in the course of time. They are extremely efficient due to their simplicity and user-friendly since they just require the user to insert coins into their pad. The vending machine for soft drinks is also becoming more increasingly popular particularly for those in the UK. This is also an easy vending machine that requires users to put just coins or money into its slot for it to open the door for your selected beverage or food item.

Automated machine for teller has been in operation for quite a while now. they are also referred to as cash dispensers or ATMs. They are extremely loved by customers since they provide them with easy accessibility to accounts. They are able to disburse any coins or notes customers like from their bank accounts however, they can also be used by any person with an account with a bank.

The first banking machine was developed in the UK and has since been adapted to different countries around the world. They are able to distribute a vast assortment of different items such as cash and bank cards.

A vending machine that is automated is a machine to offer cold and hot drinks. It operates by the consumer placing money or coins into the coin slot, after that, they place their drink of choice into the dispenser. When they place it in the machine, the drink will flow out immediately. Some vending machines with automatic technology require an exclusive cup that will let them know the kind of drink you'd like and the amount you'd like to purchase, so be sure to choose wisely prior to using the machines.

There are a variety of automated vending machines in sports venues, shops and offices. Delivery devices are very popular automated vending machines, however they're not often used to serve drinks, instead, they are used to serve warm food products.

A delivery machine that is basic is a basic machine utilized to deliver cold and hot food items to customers who want to eat in their homes. Delivery machines are becoming increasingly well-known, however, and there are a myriad of kinds, including cupcake delivery machines and delivery confectionery vending machines.

Chapter 15: Old And New Machines Machine

Pros And Pros And

A New Machine for vending Machine Pros and Pros and

Many people think that owning vending machines appears to be a simple method to earn money. But, before you decide to take the leap and invest in the machine you've always wanted it is crucial to take into consideration every aspect of the venture. The following list outlines the pros and cons of modern vending machines on the market today.

Pros

If you're beginning with no money for this type of business, modern machines are less expensive than ever before! They're also equipped with more options than ever before to provide convenience during busy times when demand is high.

In the event that you've room for them , and you have the cash to purchase them, the latest machines are fantastic! They're thinner, sleeker and more affordable than before. The latest models are also equipped with improved graphics that are more bright and easy to the eyes...it's amazing!

If you're brand new to vending, and know the basics of machines, this is a good option to get started. With modern models, it's not necessary to have any prior experience or knowledge of the maintenance of the machines. The designs are more sleek and simpler to operate as never before which makes it a fantastic opportunity to start your journey into vending.

The latest models are less expensive than before. They are available to purchase them at an reasonable cost. Even those with the highest prices cost priced at less than $200. New models are more user-friendly straight out of the box which makes them the ideal opportunity to start vending.

The new vending machines last longer than they've ever been. They are also easier to use, particularly ones with self-check-out options. These machines accept debit and credit cards making the process of buying and selling more convenient for customers.

One of the best features that's brand new to vending is the fact that there are models with self-checkout options. The customer can pay for their purchases without the assistance of an employee. This means that you can have greater customers even when you're in your computer because of a busy schedule or any other reason. As with all machines the latest machines are less expensive than they have ever been. They're still a great way to begin vending and can be bought at less than $100.

They're still an excellent way to start vending, and are available for less than $100.

The latest models are cheaper than ever. It's easy to find new machines for a reasonable price.

One of the main benefits is that the newer models include a number of components that aren't standard (meaning you'll need to locate components that are specifically designed for that model). Because so many models are on the market and replacement parts aren't always straightforward.

Cons

* The latest machines are more risky than before. Some machines can be destroyed without

warning when their self-checkout system is malfunctioning (yes it really does happen!). If you're running a tiny machine, it could be almost impossible to locate the components and then retrofit it if the machine fails.

The latest models make use of less expensive or inferior parts that are susceptible to breaking down at any moment. The new machines aren't as robust as they were and are still prone to breaking down without warning. Modern vending machines are more vulnerable to vandalism ever before.

* New machines are more prone to breaking due to their lack of durability like older models. Breaking down machines are costly and may need repairs that aren't done without prior knowledge (since the majority of companies don't offer assistance for their new machines).

The new machines aren't as sturdy as the older models. They're made of plastic parts that can be easily broken by vandalism , or normal wear and tear with time. If you have a large-volume machine, it's much more vulnerable to breakage at any time (if it's not made to handle large quantities).

* A few of these machines aren't supported at all and if something occurs, you could be in complete danger If they don't provide the repairs (or you're stuck with a malfunctioning machine). There's no uniformity on these machines, which means that they're each unique and require different skills and components to fix them.

* Along with this growth in technology is an increase in costs. This could be overwhelming for business owners who are new to the field who may not have the cash to begin with (though lower-cost models are available). They also do not last as long as older models , and cost more to fix should they break down or require repaired because of the effects of vandalism, wear and tear that happens over the course of time. The latest models are more attractive and feature better graphics as well as an improved style. It is difficult to locate an item that is identical to older models but it could be difficult to locate the latest model that's like the old one. That makes finding a new model to your current business difficult, if you are able to find any of the models.

* With the development of technology and the introduction of new features, comes an increase

in costs. This could be a problem for a business owner who is new with a limited budget to put into a vending company. The latest machines are not as similar to the older models. Getting an alternative can be difficult when you require it. The latest models have more appealing graphics and are more comfortable to the eyes...it's fantastic! But, it's difficult to locate a model that appears exactly similar to the older models.

Pros and Pros and Secondhand Vending Machine You've got an idea for a business you'd like to develop but don't know how to begin. So, you've searched on the internet for research and statistics but are you sure what these figures indicate? This is why people look for concepts through articles.

But, the majority of them are old or incorrect and are intended for professionals in the vending industry who usually get paid to create these lists. This is where this list is a good start: it lists the advantages and disadvantages of purchasing secondhand vending machines sold by vendors through Craigslist and eBay.

Pros

The used vending machines are cheaper than buying a brand new one, particularly since they are sold at an lower price. If you purchase an old vending machine the price of the machine is just one aspect of the investment you will be making. The total cost to get your business operating is more than what you'll have to pay in cash but it's still lower than the cost of buying three or four brand-new machines. Additionally, you could make money back when you decide to sell the machine later on.

* Selling machines is much easier when they have been used previously because they know that machines are able to run for hours without any problems. If the machine has not been used for a while and you wish it to be sold quickly then you'll need make minor repairs and make sure that everything is functioning properly. However, if vending machines are new, the items inside are full of air and could be sold for a price that is high.

* The machines offer superior services than they do when they're brand new, because they were extensively tested and utilized by others before you bought them so they are likely to operate

more efficiently or more efficiently than brand new ones.

Second-hand machines usually have lower problems due to being stronger than brand new ones that are made under great pressure to ensure the highest quality. It's not easy to meet the same standards, however it isn't difficult. In reality, companies who sell used vending machines are extremely careful when it comes to creating and manufacturing their products since they want their products to last and function as long as new machines. In addition, other companies are more focused on making them so that they are sold at a an affordable cost.

The used vending machines are easier to maintain and clean because they've been in use before and contain less grease or any kinds of residues inside their machines, meaning that cleaning them isn't as hard. In the end the less work you have to do, the more work you'll be able to accomplish in your business.

* The more trust people place on second-hand machines since they've been used for a long time, the more easy it is to market them since there are less issues regarding these machines. The

majority of people are familiar about these devices and are aware of what they can expect.

Cons

* You are only able to contact sellers via Craigslist or eBay that frequently use these websites, which means you'll need to look for them. Most of the time, they're not listed in other locations and therefore don't expect to do many people to contact various sellers. Furthermore, they don't offer machines through an online purchase, so you'll have to get in touch with them by email or calling them to learn more about their offerings. The seller isn't reliable and trustworthy, and does not sell second-hand vending machines, it could be difficult to secure an approval of buyers. You'll have to bargain with them and convince them that purchasing secondhand vending machines is better than purchasing new ones.

* Second-hand vending machines typically not covered by warranties since they've been in use before. The reason for this is that sellers rarely intend to offer assistance or assistance when selling their goods since they're too expensive and difficult to sell.

* If you plan to sell old vending machines after you have purchased the machines, it's beneficial to include a warranty in your purchase contract. In this way, you will safeguard buyers from potential loss and make sure they feel secure buying your products.

* Vending machines that are used are difficult to sell as they're usually sold without the warranties or support of the manufacturer who originally made the machines. This is one of the biggest hurdles that you must face when you want your business to be successful. Prior to selling any equipment, conduct some research to find out which machines provide the best support and warranty, so that your machine stands out from the rest.

* The items you'll sell through vending machines may need to be repackaged or verified to ensure that they're still fresh prior to selling them. So, you'll have spend your time on these small tasks which aren't important when performing other tasks.

* A machine damaged in any way doesn't carry any value and buyers aren't likely to buy it even if they require the compressor or another

component that the unit. Don't expect too much from selling an item that is secondhand and comes without the original packaging.

If you're considering buying second-hand vending equipment, look for dents or scratches on the exterior. If everything is in good shape the machine should be reliable and stable and if it's not, the seller may have a tough to sell it.

* With the aid of used vending machines, it is possible to can generate a significant profit for your business by offering top quality items at reasonable prices. But, you must be aware when looking for them and know more about them prior to purchasing them. This will enable you to locate a reliable seller, and also ensure that your products are shipped in good condition so that they can be sold quickly and without any problems. If you are able to market used vending machines efficiently The marketability that your products will improve which can lead to more sales and profit for you as well as your company.

Chapter 16: What Is The Reason To Should You

Invest In A Vending Business?

You may have guessed by the title that the book is going to be focused on vending machines! The book will discuss how having vending machine businesses can be a profitable investment. This will include information on the workings of vending machines and what type of investment is required and the reasons to think about purchasing one!

Take this book as your guide for investing into a vending machine company. This chapter will start at the simplest level and then move to the next level. When you're done with this section, you'll have everything you need to know about how to get started in the business of owning the vending machines of your own business.

What makes vending machines appealing as a way to invest?

The easiest answer is profits and the types of cash-making that is involved. In the end, you don't require a large capital investment upfront to begin your own business, do you? It is possible to start at a low level and grow as you go. This isn't something that can be done in a single day however, there are important advantages! These are the benefits:

* Costs for starting are low. Vending machines require a minimal investment to start. It is possible to purchase and store them with goods between $300 and 2000 dollars which makes it an economical business option when compared with other options.

* Low costs for overheads. Apart from your initial investment and inventory, you do not need to incur any costs. You won't even require any additional employees - you could manage your business by yourself! This means that you pay very little in terms of rental and other utilities.

• Passive earnings. The vending machine can earn you cash while you're sleeping, at work or just doing your routine. How awesome is this? You're always earning cash for you, in a way that is automatic!

* Flexible schedule. Contrary to other business concepts vending machines are accessible all

hours of the day, so you have the option of deciding when you want to plan your work around them (within the reason).

* Good products. Your products are regularly used, not like other. Nobody will purchase chocolate bars even if they aren't eating it in the moment!

Three Ways to Earn Money with Vending Machines

The most appealing aspect of owning vending machine company is the numerous opportunities to earn. There are three primary reasons people choose to invest in the operation and management of vending machines:

* You can sell them. This is the easiest method to earn cash. All you need to do is purchase the equipment and stock them up and profit by selling them! Many people opt for this method since it's less maintenance-intensive and the majority of products only need replacement every now and then.

* You can make use of affiliate hyperlinks. If you're giving away sweets and drinks you keep in your vending machine then you'll make money by placing affiliate links on the shelves of the general

public. A third party to manage all of this is the best method to start.

* You can sell your products with greater profit margins. Machines for vending are better than all other kinds of business. The profit margin for vending machines of 40% which means If you think about it this way, any item with a higher margin of profit is a better fit in terms of your financial budget.

The Best Investment You Could Make

Running a vending machine company is among the most lucrative ways to earn passive income can be earned. Since it's fully automated, you won't need to think about anything else except making sure your machine is properly filled and operating. These machines are there to stay for the long haul and you don't need continuous attention to ensure they're operating. They're selling goods. Every person wants to earn a steady income, especially when they are looking to pay for their costs on the side, while working full-time.

Conclusion

The idea of starting your own business may appear daunting It's not, but it's worth it at the end of the day. You'll be able to earn money efficiently and profitably by focusing on something you are familiar with instead of something you're not interested in similar to many other occupations.

The idea of starting a vending machine business is also a great idea since you do not have to handle customers in person , or handle employees or employees. It is easy to relax and watch your earnings grow without worrying about any aspect.

You'll quickly understand what you have to do and what you should purchase. You'll also discover how much items cost, making it much simpler for you when it's time to determine what number of machines to purchase.

There's some danger involved in starting vending machine companies and that's the reason that can make the money so much. In a vending company it is possible to utilize various payment methods.

If you'd like to go the safer option, you can accept only cash or credit cards. If you're keen to try something new, place a coin machine on the machines and let people to pay using coins. A lot of people choose this method because the price is lower. It's your choice which payment method you prefer to choose.

Additionally this is also beneficial since if your machine fails or is destroyed by another party because of a reason or another, no one will pursue your property for the lost money. Additionally, you're the proprietor of your own company, which means that you do not have to answer to anyone. You're not a worker therefore you don't have to be concerned about your work. Only one person to answer for yourself is you which is one thing you should avoid.

www.ingramcontent.com/pod-product-compliance
Lightning Source LLC
Chambersburg PA
CBHW071223210326
41597CB00016B/1922